Dark Matter, Neutrinos And The Big Bang.

The Nucleon-Deuteron model 2017
&
'Atomic Replication' Theory

Created & Developed by Linda Hutchinson.
Illustrated by Bethany Hillier
Mathematics Consultant: Charlie Hillier.

This book is dedicated in loving memory to my Beautiful Mother, Diana Hutchinson.
Who's bravery inspired me to search for the true nature of existence and the reality of Space-Time.

Diana Hutchinson.
11 October 1956- 2 March 2009.

Using This Book.

This book contains the Nucleon-Deuteron Model & 'Atomic Replication Theory', with a full explanation of how it was developed and how it works.

The book begins by explaining sciense's current standard model, then goes on to explain the sub-atomic parts of the Nucleon-Deuteron Model 2017.
It goes on to explaine the key information that was relevant to the creation of the Model & Theroy.

The book takes us through the process of The 'Big Bang' & explaines what happened 'before' the 'Big Bang', during Cosmology's first second. Cosmology's first second is broken down into Epochs or Eras to help us understand the events which took place.
Each chapter is divided into Eras or Epochs. Each Epoch scientifically explaines Force and Particle production that should have taken place during that time.

Each chapter gives scientific evidence & Equations to explain the models findings.

The Model Begins with the Singularity and how & why it should have developed into a whole universe. The Model describes how the singularity particle should have caused the first Forces to form as a natural bi-product of cause & effect, due to the laws of Physics.
It gives a step by step guide to how the first atom should have been formed, revealing for the first time, the anatomy of the atomic structure in great detail. Answering many of physics previously unanswered questions along the way.
The book takes us through the entire process to Galaxy formation.

The Classic Model of the Atom.

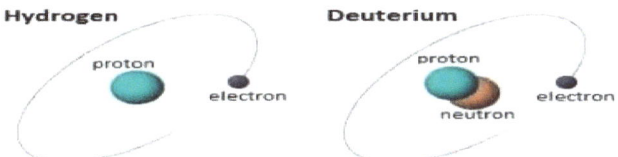

Diagram showing the Standard Model Atomic structure of Hydrogen & it's Isotope, Deuterium atoms.

We know that all matter in our universe is made up of Atoms. There are different types of atoms. These different types make up the Chemical elements ie Calcium, Iron, Gold etc.
Each atom in a Bar of Gold is identical to the next.

Atoms are made from even smaller particles, called Sub-atomic particles. The main types of Sub-atomic particles are Protons, Neutrons & Electrons. An atoms type depends on how many of these main Sub-atomic particles it has.
The simplest and first atom to have existed in the early universe was the Hydrogen atom.

A single Hydrogen atom contains 1 Proton, 1 Neutron & 1 Electron. The Positive Proton has an electrical charge & the Negative Neutron has no electric charge. This similar to a battery. Together the Proton & Neutron make up the centre Nucleus of the atom. The Electron is smaller and is formed by the Nucleus and exits orbiting the nucleus in shells (orbital levels) Almost like a tiny solar system. An atom will normally have the same amount of Electrons as Protons. An Electron is negative and carries a negative charge.

Deuterium is an Isotope of Hydrogen because it still only has 1 Proton, but unlike Hydrogen it can hang on to it's Neutron. It was developed naturally from Hydrogen and is the stepping stone from Hydrogen to Helium production.

We know that the Nuclei's Protons & Neutrons are made of even smaller Sub-atomic partcles.

Our understanding of the atomic structure has had its limits. Atoms are so small that even the most powerful of microscopes can't see directly into the structure of the Protons & Neutrons. To gain a better understanding of the Anatomy of the atom, Scientists smash Protons, Neutrons and Electrons together to see what smaller Sub-atomic particles come out of the collisions. They have found that Neutrons & Protons also conatain Quarks, Gluons and Bosons. They also found that energy would also be released in the form of sub-atomic Neutrinos. Science is now trying to find out how these smaller sub-atomic particles piece together into a working structure of the atom.

The Nucleon-Deuteron Model 2017 is the first of its kind to show , in great detail, how these paritcles should form a simple but highly organised structure.

Identifying The SubAtomic Parts Of The Nucleon-Deuteron Model 2017 Diagram.

The Diagram shows the Nucleon-Deuteron Model as it should be seen inside the **Deuteron** of the Deuterium Atom. Depicting a simple but elegant **spiral structure of both Quark, Neutron & Proton.**
As we know the universe is thought to have began with the **singularity,** which can be thought of as an **Anapole Majorana particle.**

- This Model agrees with the idea that the singularity/ Majorana should have experienced a change of state during a heat sink process of the 2nd law of Thermodynamics. This should have caused an energy release in the form of Neutrinos.

Therefore this Model & Theory begins by **postulating the existence of primitive (Singularity) 'Majorana Neutrinos'** during cosmology's first second. Existing during Electro-Weak era/epoch at a **Temperature of 1.8125×10^{21} Degrees Kelvin. It's Speed 2.9979×10^{8} mps, Frequency of 1.8737×10^{32} HZ at a Wavelength of 1.6000×10^{-24} mm (yoctometre).**

- **A left handed Majorana Neutrino** should also have it's **entangled right handed Majorana Anti-Neutrino counterpart**, the Model identifies it's properties.
- **A Majorana Neutrino & it's Anti-Neutrino** seem to cause **W-, W+ and Z Bosons** to form between themselves. Creating a **flux tubular** body structure, and the first Forces.
- Two of these Boson structures form a **Quark**.
- The interactions between the W- and W+ Bosons & their associated Neutrinos, seem to **determine the Flavor & Color of the Quark.**

- The configuration of Up and Down Quarks determine the type of Hadron which is created. Either a Neutron or Proton.
The Neutron contains 3 Quarks: Down Quark, Up Quark, Down Quark.
The Proton contains 3 Quarks: Up Quark, Down Quark, Up Quark.
The division between the Neutron & Proton lies between Quark 3 & Quark 4.

In The Nucleon-Deuteron Model:

- **Coloured spheres** depict Majorana Neutrino/ Anti-Neutrinos. Paired by adjoining W & Z Boson/Gluon Fluxtubular fields.

- **Red lines** depict the W- Boson/ Gluon field. Always coupled with a Green W+ Boson /Gluon field.

- **Green lines** depict W+ Boson/Gluon field. Always coupled with a W- Boson/Gluon field.

- **When 2 Neutrino/Anti-Neutrino pairs & their 2 Boson/Gluon fields meet:**
  they should create the Electro-Magnetic force. This should happen as a result of the famous Neutrino flip. This action caused the first Quark to form. Although there are 4 Majorana Neutrino heads(+) & Anti-Neurtino tails (-) present only 3 are used to form each Quark. The remaining 'Lonley' Neutrino/Anti-Neutrino is used as part of the next forming Quark. The 'Lonley' (+) head or (-) tail and it's counterpart are depicted in Blue. Their Boson/Gluon field can be either W- or W+. They are depicted as blue because they are 'connected' to the Red W- & Green W+ pair via **Electro-Magnetic force.**

- **Yellow lines** depict the **Higgs Boson** Field. The Higgs has different functions to the W & Z Bosons. It does more than just add mass, and was created in a specific way. The model expalines how & why the Higgs should form. The colours surrounding the yellow Higgs lines indicate the Quark color and flavor which the Higgs Field is directly involved with.

- **Black lines** depict **Z Bosons**. A **vacuum** space, which serve as tiny **Event Horizion** points between the W Bosons. The Z Boson seems to facilitate all manner of functions, which directly effect Beta decay processes.

The Model finds Hydrogen & it's Isotope Deuterium's structure to be spiraled as Quarks seem to form in a spiral formation. Spirals are found everywhere in nature, from tiny fossils to Galaxies. We find a similar more advanced spiral structure in RNA & DNA. It seems that the spiral structure pattern which governs our universe might have began with the spiraling formation of Quarks at $10\wedge12$ seconds.

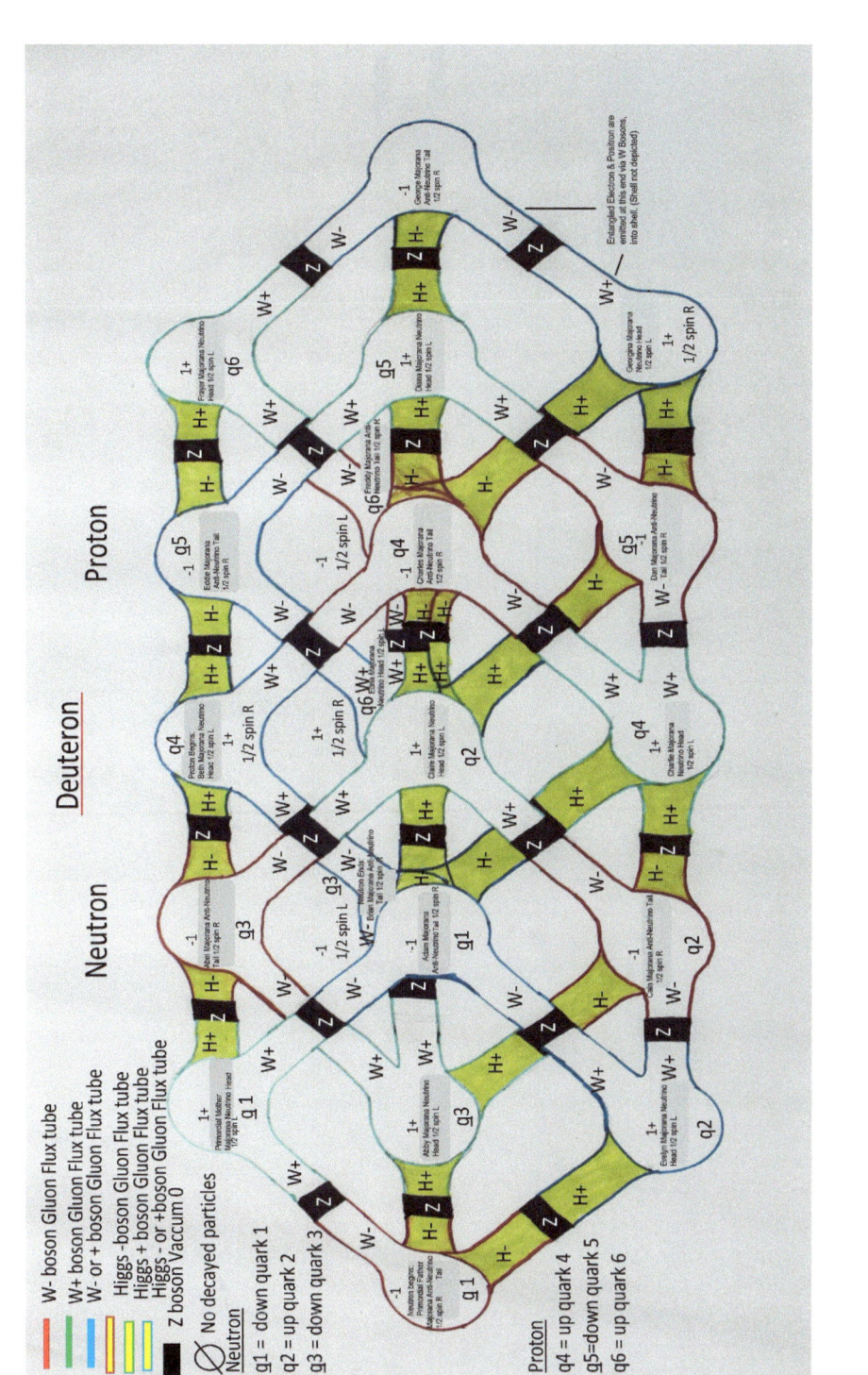

Introduction
To The Nucleon-Deuteron Model

The Nucleon-Deuteron Model is based on how naturally occurring isotopes of Hydrogen are produced. The model is built upon the principle that particles have Anti-particles. The model shows how each subatomic particle should have developed as a result of cause and effect. The development of these particles is placed in the envirienment of the early universe during cosmology's first second of the "Big Bang." The process begins with the production of the 1st Anapole Majorana particle and takes us through Force & Particle production, showing the Anatomy of the Atom.
It explains how Hydrogen and Deuterium form begining with the Majorana singularity Anapole point which should emit Majorana Neutrinos, how Majorana Neutrinos are a dipole structure which seem to cause Electro-magnetism and W+, W- and Z Bosons to form. It explains how this event itself should have caused Quark formation.

It details the inner mechanics of the atomic structure and how Neutrinos, Bosons & Quarks work together as a simple but highly organised mechanism. It shows how this mechanism should have created the Neutron & Proton, which naturally developed as a result of cause and effect.

The anatomy of the atomic structure seems to reveal the structure and nature of Space-Time and Dark Matter & Dark Energy.

How the Nucleon-Deuteron Model 2017 & Atomic 'Replication' Theory was Developed

The Nucleon-Deuteron Model begins with the formation of the Majorana type singularity particle. Marjorana particles were theorized by British scientist Paul Dirac in 1928. He developed the Dirac equation which says that particles have Anti-particles. His equations show that particles should have opposite Chirality quantum properties to their Anti-particles. A Majorana particle is its own Anti-particle and thus is both plus and negative simultaniously.

The singularity is thought to have been the very first state of existence of the universe, which existed around the time of the Big Bang. It is thought that this singularity particle was its own Aniti-particle, therefore it was a Majorana singularity particle. We also Know that a Majorana particle has an Anapole 'Do-nut' structure.

11
Neutrinos.

In 1930 Austrian-Swis scientist Wolfgang Pauli postulated the existence of Neutrinos. He was nominated by Albert Einstein for the Nobel Prize in Physics, which he received in 1945 for the Exclusion or Pauli principle.
Neutrinos are tiny points of energy which are produced during particle decay. They seem to be part of the process which ensures energy conservation in particle interaction. Energy conservation laws say that energy cannot be destroyed and must be conserved during particle / energy interactions. When a particle anhilates, or changes it's state of being, some energy is released and is seen as a different particle, the Neutrino.

There are three types of Neutrino that have been found so far. Each Neutrino is named after the particle which causes its existence. These are the Electron Neutrino, Muon Neutrino and Tau Neutrino.

The Nucleon-Deuteron model suggests that the Majorana Neutrino & it's Anti-Neutrino counterpart should have formed as a result of the heat sink process of the 2nd law of Thermodynamics, which caused a decay/change of state of the Anapole Majorana singularity particle.

Paul Dirac predicted that the Neutrino must be left-handed and have a right-handed Neutrino counterpart. He also predicted that the Anti-Neutrino must be right-handed and have a left-handed Anti-Neutrino counter part. This makes four types of Neutrino.

In 1937 Italian Physicist Ettore Majorana theorized that if these Neutrinos had no mass there would be no need for there to be 4 types of Neutrino. There would only be a need for left-handed Neutrinos and a right- handed Anti-Neutrino counterparts. He also theorized that it would make no difference if they had a small mass because the Neutrinos might be able to change state from left to right-handedness, this is now known as Neutrino colour or flavor.
The new Nucleon-Deuteron Model agrees with Ettore Majorana's theory that both Neutrino & Anti-Neutrino can have a small mass.
This fact that Neutrinos have a small mass has been proved by experiments that were conducted in 1998. They were called the "Superkamiokande & SNO experiments". The discovery won the 2015 Nobel prize. The Nucleon-Deuteron model also finds that Neutrinos seem to change state or flavor as Ettore predicted.

12
Parity Violation

As a fundamental law, lepton numbers should be conserved, staying the same before and after interactions. Science says that in the early universe the Anti-matter and normal Matter should have annihilated each other leaving just pure energy. However there is a violation of this idea, parity symmetry violation says that the ratio of Matter and Dark Matter were once the same in the early universe. An event happened which tipped the parity symmetry causing a-symmetry. Dark Matter became more abundant in the universe than Matter. The Nucleon-Deuteron model reveals how and why this parity symmetry violation should have happened.

The Majorana was a 'do-nut'!

An article entitled "Anapole Dark Matter" was published in Journal Physics Letters B. By Professor Robert Scherrer & Chiu Man Ho from the Vanderbilt university.
They found that the Majorana particle had a rare form of Electro-Magnetism, which created a 'do-nut' shaped Electro-Magnetic type field. This field was an Anapole field. They suggest that the anapole field was generated by a toroidal electrical current and that the field itself was confined within the torus region. It was electrically neutral. The anapole current is different from the field created by a Dipole structure, seen in normal Electro-Magnetism. A normal Electrical field spreads out and can interact with Electro-Magnetic fields whilst moving or stationary.

The Anapole does not interact with other fields while it is still, it must be moving at a high velocity. It causes its own unique energy. It is thought that this was the state of the Majorana particles in the early universe. Taking this information into account, Along with the findings from recent scientific observations of black holes. It appears that the Majorana particle has a lot more than we thought in common with Black holes.

Diagram of an Anapole field

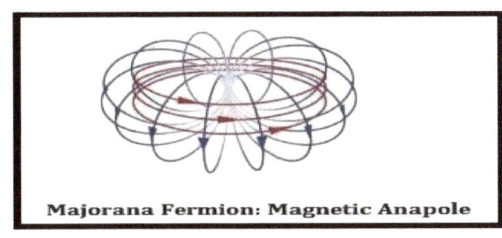

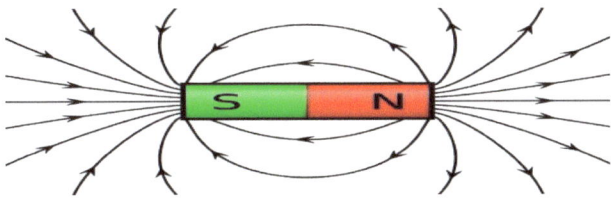

Diagram of a Dipole Magnetic field

The Majorana particles in do-nut form seem to resemble black holes but on a micro scale. To find out if the two structures had any other similarities it was necessary to research black holes. It was observed that black holes seemed to work in a very similar way to the Majorana.

Quarks, Neutrinos, Bosons And Flux Tubes In The Nucleon-Deuteron Model

A paper was recently released by Derek B. Leinwber from the Centre for the Subatomic Structure of Matter, department of Physics, University of Adalaide Australia. The paper was entitled "visulaisations of QCD". The paper explained how the gluon structure which holds Quarks together seems to be a flux tube. It shows the density of the vacuum action of the flux tube as it gradually reduces. It was stated that energy is required to expel the vacuum where a linier confinmen potential is felt between Quarks. It showed that these flux tubes retain their diameter when they stretch with distence.

Derek B. Leinwber's work led me to infer that these vacuum areas might act like micro black holes and that they must facilitate all manner of processes. The information also led me to realise that these gluon flux tubes which hold Quarks together might also be responsible for holding Majorana Neutrinos together on a much smaller scale. It became apparent that the gluon flux tubes must have originated from the production of Majorana Neutrinos. Therefore the information obtained from the article by Derek B. Leinwber on gluons was taken into account and applied to Majorana Neutrinos at a much smaller subatomic level.

Derek B. Leinweber's model was a computer generated animation created by the centre of Subatomic Matter to show the gluon flux tube action.
 The simulation shows how the Gluon structure seems to be in the form of flux tubes. It shows Quarks moving around each other in a gluon field, which was cube-like in shape. The Quarks did not seem to create a definate stable structure but seemed to move freely. This theory raised some interesting questions that the Nucleon-Deuteron model seems to be able to answer. Therefore the Nucleon-Deuteron model owes much credit to those brilliant scientists who discovered that gluons are flux tubes.

The Nucleon-Deuteron model creates a cube-like structure of the Hydrogen Nucleon and the Deuteron. It is created by the spiraling structure of the forming Quarks. It details the mechanical interactions which form the strong structures. It shows that Majorana Neutrinos and the gluon field are not seperate but combine to form an entangled Neutrino & Anti-Neutrino particle pair which share a body structure. These entangled bodies form Quarks. The Nucleon-Deuteron model shows how they seem to form and reproduce, and how each Boson serves a purpose.

Scientists now think that Neutrinos are Majorana particles. With this information It was possible to theorise how a singularity Majorana point could turn or decay into a Majorana Neutrino. As we know the Majorana Neutrino must contain within it, its own Anti-particle. It seems have a Dipole shape, which "evolved" from an Anapole initial state. The Nucleon-Deuteron model consideres the Majorana Neutrino as a positive & negative entangled pair. Their structure seems to cause the Gravitational & Electro-weak force. This led to a realization that the pair must be entangled via the gluon flux tube structure and that the dipole Neutriono heads and Anti-Neutrino tails were responsible for the flavors or states of the Quark.

From this information it was possible to theorise that if Majorana Neutrinos create Gluons, they must some how still be an active part of the Gluon structure. It was possible to create a model where the dipole Majorana Neutrino heads and Anti-Neutrino tails themselves formed the Quark up and down states. The Nucleon-Deuteron model was able to show the Neutrino, gluon/ flux tube, Quark structure to be a stable model which explains all interactions due to cause and effect. It was possible to infer that since Marjorana Neutrinos seem to make up Quarks, the Quarks might behave in a similar way because they have "inherited", or been "programmed" or "remembered" their behaviour from their parent Majorana Neutrino and Majorana singularity states. This could be evidence of particle "Evolution".

The following questions were raised at this point. If a Neutrino contains a micro black hole at its centre point, does it draw anything into itself? If so could this be decaying matter? If matter decayed through these tiny holes where did it go? As we know energy cannot be destroyed but merely changes state. In order to answer these questions it was necessary to include Beta Minus, Beta Plus and Alpha Decay. Which led to more information being found concerning these processes.

During the construction of the model, the appearance of a checkerboard pattern made it clear that there seeemed to be a divide between the Majorana Neutrino dipole heads and Anti- Majorana Neutrino tail sections of the body. These zones were linked at separation points. Which seem to be Z Bosons. The cube revealed oscillations between the higher (+) states and the lower (-) states of the Majorana Neutrino dipole heads and Anti-Majorana Neutrino tails, when the Elecrto-Magnetic /formation/ energy passed through each "Neutrino / Anti-Neutrino pair".

16
What The Nucleon-Deuteron Model Should Find

This Nucleon-Deuteron Model finds Answers to many previously unanswered fundamental questions of Quantum Physics.
The model walks us through what is found to be a primitive particle 'Evolutionary & Replication' process which allows for the singularity/Majorana to naturally create the known forces in turn as products of the laws of cause & effect.

- It shows how the singularity/ Anapole Majorana particle should have created primitive Neutrino & Anti-Neutrino pairs due a heat sink processes of the 2nd law of Thermodynamics which causes temperature fluctuations of the Anapole Majorana particle. At a temperature of 1.8125×10^{21} degrees kelvin the fluctuation should effect the wavelength of the particle until it 'evolves' to the wavelength of 10^{-24}mm which is a yoctometre, the size of a Neutrino. With this we also have its Anti-Neutrino. This is not the same as our normal Electron Neutrino, Tau Neutrino or Muon Neutrino. It is far more primitive and seems to have been formed directly from the singularity/ Majorana particle as a natural means of energy conservation.
- To differentiate between this early type of Neutrino and its later cousins, the Nucleon-Deuteron model & Theory refer to the subatomic particle as the Majorana Neutrino & it's Anti-Majorana Neutrino. Their Neutrino pair speed was calculated to be 2.9979×10^{8}mps. The same speed was found for the Majorana/ Singularity particle. These Speeds are of importance because they show that we can use the Majorana particle and it's primitive Neutrino/Anti-Neutrino pairs in our calculations rather than using light speed which is 2.9979×10^{8}mps (short version). The reason for doing this is that we know that Photons which make up light were not yet formed at this point in the early universe. Therefore Photons were not present for their energy to be utilized.
- The model offers an alternative to using the Photons properties for these equations. The model uses the Majorana & it's Neutrino / Anti-Neutrino speeds and properties for equations which have uncannily similar properties to the later Photon. The model finds why their properties are so similar. We can also postulate from this that Neutrino & Anti-Neutrino pairs can exist in both particle and waveform as it's descendant the Photon does.
- The model finds that during GUT era/epoch the Neutrino / Anti-Neutrino pairs seem to work together due to natural laws causing the

- creation of not only the Known Forces but also the W+,W- & Z Bosons & Quarks.
- The model details how the Quark system should be a highly organised mechanism which seems able to 'replicate'. This action seems to create the first primitive Neutrons & Protons. It shows how the process should continue until the Hydrogen Atom is complete.
- The model goes on to show Electron & Positron formation. Then Photon & Anti-Photon formation.
- The Model Explains the Higgs Boson & Field formation, how and why the Higgs is formed and identifies more of its functions & where it should lie in the atomic structure. The model shows how the Higgs should be effected during nuclear fusion, giving more information on binding energy and mass defect.
- The model shows how Quark color and flavor change should be a direct effect of Neutrino interaction.
- But by far the most fun thing that the model seems to have found is the identity of Dark Matter & Dark Energy.
- The model seems to show how and why Dark Matter & Dark Energy were naturally produced as a product of cause and effect.
- It shows how Dark Matter & Dark Energy seem to have began to develop in the very early stages of the universe during GUT era.
- It Seems to find why they are almost undetectable to our instruments and how and why they should have a gravitational effect on normal matter.
- The model seems to find the reason why there is more Dark Matter & Dark Energy in the universe than normal Matter & Energy. It gives an explanation as to how the amount of Dark Matter production in the early universe seems to have overthrown the production of normal Matter, leading to there being around 5% normal Matter, 27% Dark Matter & 68% Dark Energy.
- The Model seems to pin point the overthrow to approx 10^{-4} Seconds at a temperature of approx 10^{12} degrees kelvin, just after Hadron era/epoch began. The model finds it due to a natural process of Beta minus decay of the Neutron. The model finds this point to be the earliest evidence of Parity Violation in the early universe.
- The Model & Theory seems to find the cause of quantum Entanglement "Spooky action at a distance" and attempts to explain how it should work.
- The model seems to find evidence for particle 'Evolution & Replication' & ' Innate particle memory' or 'programming', leading eventually to RNA & DNA formation.
- The Nucleon-Deuteron model reveals a Micro Black Hole in the Majorana Neutrino state.

- The Nucleon-Deuteron model shows how the Photon & its Anti-Photon, seem to be an evolved state of the Majorana with the innate / programmed ability to experience "states" similar to its Majorana ancestor. The Photon seems to retain its " memory" of all states of its evolution to be later called upon in order to display qualities of the Majorana.
- The model also depicts how Space-Time seems to have formed and can bridge the gap between the laws of Classical Physics & Quantum Physics.

- The Nucleon-Deuteron model shows evidence that another previously undiscovered dimension might lie within its structure.
- The Nucleon-Deuteron Model reveals evidence to support the existence of a singular dimension, which lies parralel to our own. It reveals how, where and why it should have been formed and its quantum properties.

- The Nucleon-Deuteron Model supports recent theoies of the existence of an Anti-Photon.
- The Nucleon-Deuteron model also suggests that the Majorana particle was not only Do-nut shaped but that it may have been a polarized bubble type of structure. The model shows how it seems to have "evolved" into an entangled Majorana Neutrino/ Ani- Neutrino. Rather than being annihilated or destroyed after emitting a Neutrino & Anti-Neutrino.
- The Nucleon-Deuteron model shows evidence to support the hypothesis that the Neutrino & Anti-Neutrino may be an "androgenous" head & entangled tail structure. There is evidence to support that the Majorana structure also held the blueprint for the male & female species of living organisms.
- The theory discuses if the consciousness survives physical death and if so what happens to it and its state of existence and intelligence. It also investigates how this relates to the Nucleon-Deuteron model.

How The Nucleon-Deuteron Model works
Method

Neutrinos:
We know that two Gluons form a Quark. We know that three Quarks usually form Neutrons and Protons. Neutrinos seem to form as a product of decay. For example, Electron Neutrinos are produced due to electron decay. Neurtinos appear as different types & are named according to the particle that they formed from.

Types of Neutrino:
Electron Neutrino & its Anti-particle the Ani-Electron Neutrino.
Muon Neutrino & Anti Muon Neutrino
Tau Neutrino & Anti Tau Neutrino

The Nucleon-Deuteron model sugests that there must be a Majorana Neurtino and it's Majorana Anti- Neutrino.

When a Majorana particle anhilates or changes form, it should emit some energy, the new energy should be similar to its parents initial Majorana state. The new energy should rapidly develop an anapole structure, then stretch like its parent Majorana into a dipole structure, due to the heatsink process.

The wavelength of the new dipole structure is in the order of $x10^{-24}m$ which is equil to a yoctometre. The yoctometre is in the order of the Neutrino. The new dipole energy becomes a Neutrino / Anti-Neutrino entangled, dipole pair. These particles seem to Create the Gravitational force, Electro-magnetism, Nuclear strong force, Neuclear weak force, Bosons, Quarks, Neutrons, Protons and Electrons and Photons.

The development of these particles is placed in the enviornment of the early universe during the first second of the "Big Bang" it shows the anatomy of the atom.

The anatomy of the atomic structure seems to reveal the structure and nature of time-space and Dark matter & Energy.

From The Beginning...
Plank Era/Epoch
In The Nucleon-Deuteron Model

A lot happens within the first second of creation. Cosmology breaks this first second down into Eras/Epochs to help us understand the events which caused our universe to develop the way that it did. This model takes us through the first second.

Plank Era/Epoch is the earliest & smallest amount of time that could have existed in the universe. It marks the very beginning of time itself. Plank Era/Epoch existed at 1.35×10^{-43} seconds.
The Temperature was thought to have been between 10^{32} Degrees Kelvin to 10^{32} Kelvin.
The Nucleon-Deuteron model gives a temperature of 1.8125×10^{32} degrees Kelvin. The universe is thought to have started out as a tiny point at the smallest length possible, this is known as Plank length. This is 1.6000×10^{-35} mm. Which is the tiniest fraction of a millimetre possible.
To calculate the temperature of an Epoch we use wiens law over the wave length. The length of the universe at this time is Plank length and so the wavelength at this time is also plank length. Therefore the equation is as follows:

$$T = \frac{0.0029}{1.6000 \times 10^{-35}}$$

$T = 1.8125 \times 10^{32}$ degrees Kelvin.

Most scientists agree that the universe must have began as a singularity. A tiny point in a hot vacuum. This point has been described as a Majorana type particle. It is thought that this single point held within it, the energy potential which created all matter & energy that our universe is made from. The Majorana singularity point must have held within it both positive & negative energy potentials. This point is thought to have expanded to produce the forces that govern our universe:
Gravational Force
Electro-magnetic Force
Nuclear Weak Force
Nuclear Strong Force
It is thought that from these forces Quarks, Neutrons & Protons form to make Atoms. The process begins with the Majorana singularity point which must have stretched due to the heat sink process of Thermodynamics.

The second law of Thermodynamics says that you can not take heat from something (in this case the Majorana singularity particle) and turn it all into

mechanical work. The law says that some heat has to pass through a "heat sink" which is a lower chamber or area that has a lower temperature.
The amount of energy that comes out is determined by the difference between the source and the sink. During this process some heat or energy gets lost as the heat source and the heat sink equilibrize. known as entropy. The law explains how the Majorana singularity point experienced its first change of state which brought about the beginning of the universe.

The idea is that the Majorana singularity particle experiences a change of temperature at some point of its "body". This temperature difference between the effected part and the unaffected part cause a stretching of the particle. The temperature change might have come about due to a vibration of the particle.

We know that Electrons and Photons were not yet in existence during the heat sink process of the Majorana singularity point. Therfore Electron Neutrinos and their cousins would not be able to form at this point.

The particles that were emitted would naturally be Majorana Neutrinos because these early Neutrinos should be produced in the "anhilation" or heat sink process of the Majorana singulaity point. Therefore to distinguish these early Neutrinos from their cousins, they shall be refered to as the Majorana Neutrino and Majorana Anti-Neutrinos for the purposes of this Model.

Throughout the Nucleon-Deuteron model it is revealed that Neutrinos, Anti-Neutrinos, Quarks, Neutrons & Protons also seem to experience a similar heat sink or energy difference causing a slight imbalance which allows polarity and energy to manifest. This process seems to give rise to the very first oscillations of the universe.

How The Singularity Began To 'Evolve'

Majorana singularity particle as a vacuum point.

To begin with, the model shows how an anapole Majorana Singularity particle experiences the "heat sink" process. This begins at the stage where the particle is still a neutral point. The cause is thought to be vibration. The temperature difference caused by the process causes the Majorana point to expand.

This process would have happened at Plank time: $<10^{-43}$ seconds.

Cosmology's standard model shows that at Plank Time the universe was at least 10^{32} Degrees Kelvin.
The Nucleon-Deuteron model finds that as Plank epoch phased into GUT epoch the temperature cooled or began to equibralise from 1.8125×10^{32} degrees Kelvin to 1.8125×10^{27} degrees kelvin by the end of the epoch. As this happened the wavlength of the particle/universe stretched.
During the Plank Epoch, cosmology says that all of the forces that govern the universe were compressed into this tiny Majorana singularity point. The Gravational force, Electro-magnetic force, Nuclear Weak foece & Nuclear Strong force. It is thought that together they created the combind Superforce.

However the Nucleon-Deuteron model shows that these forces might not have existed ready made and compressed inside a tiny point, they seem to have each been created as part of a cause and effect chain reaction at different points during the first second. This is discussed furthur during the Gut and Electro-Weak Epochs.

Summery Of The Equations For Plank Era/Epoch

Equations for Plank epoch:
C= speed of light 2.99792458×10^8 (metres per sec)
MSS = Majorana singularity speed/ velocity 2.9979×10^8 m/sec
$\sqrt{Gh}$= Plank time $<1.35 \times 10^{-43}$ (seconds)
pl = Plank length 1.6000×10^{-35} (mm)
h= Planks constant
T= temperature (Degrees Kelvin)
t= time (seconds)
λ= wavelength (mm)
f= frequencey (HZ)
E= energy (Joules or Gigga electron volts)
v= velocity (metres per sec)

To find the temperature at Plank epoch:
We know the wavelength is equil to plank length 1.6000×10^{-35} mm. Using wiens law we can say

$$T = \frac{0.0029}{\lambda \text{ or pl } 1.6000 \times 10^{-35} \text{mm}}$$

Temperature = 1.8125×10^{32} k

To find the frequencey at Plank epoch:
For this first equation we will use the speed of light to deturmine the frequencey because this is how it is traditionaly done. We need the frequencey of the Majorana singulaity particle to deturmine its velocity. This equation can also be done using the velocity of the Majorana singulaity itself, which we will look at once it is found using these equations.

$$F = \frac{C}{\lambda \ 1.6000 \times 10^{-35} \text{mm}}$$

Frequencey = 1.8737×10^{43} HZ

Now we have the frequencey of the Majorana singulaity we can work out its velocity by multiplying its wavelength by its frequencey.

V= λ 1.6000×10^{-35} mm x f1.8737×10^{43} HZ
velocity = 2.9979×10^8 m/per/sec

We can see that the velocity of the Majorana singulaity particle is very similar to the speed of a Photon. This is why it is easy to use the properties of the Photon for most of the equations. The only problem with doing so is that we know that the Photon did not yet exist at this early stage and it also has a different wavelength & properties to the Majorana singularity which will effect calculations.

If we go back to our earlier equation to find the frequencey of the plank epoch & Majorana singulaity particle, we can use its own velocity to solve this.

$$F = MSS$$
λ or pl 1.6000×10^{-35} mm
frequencey $= 1.8737 \times 10^{43}$ HZ

We can see that the result is the same. It is possible to use the Majorana singulaity particles properties to calculate equations leading up to the production of the Photon. To find the Energy of the Plank epoch & the Majorana singulaity particle: Traditionally plank energy Ep is 1.96×10^{9}J the Nucleon-Deuteron model finds the energy at this point to have been 1.2291×10^{10}J.

h x MSS 2.9979×10^{8}
λ 1.6000×10^{-35} mm
Energy $= 1.2415 \times 10^{10}$J
or

E= h x f1.8737×10^{43}HZ
Energy $= 1.2415 \times 10^{10}$J

To find the mass of the Majorana we should use its own speed.

The formula is: Mass (m)= Energy (E)
    Majorana speed² (msp²)
so
Majorana Mass = Majorana Energy 1.2415×10^{10}J
    MSS $2.9979 \times 10^{8^2}$ m/sec
 = Majorana Mass: 1.3814×10^{-7}kg

Because E=MC²

E= 1.3814×10^{-7}kg x $2.9979 \times 10^{8^2}$ m/sec
    = 1.2415×10^{10}J

Gut Era/Epoch

The Majorana particle vacuum expanding.

Plank epoch ended at 10^{-43} sec.
Gut epoch began at 10^{-43} sec and lasted until to 10^{-36} sec. At the beginning of GUT (Grand Unified Theory) epoch the Temperature was 10^{32} degrees Kelvin and dropped to 10^{27} degrees Kelvin by the end of the epoch.
Cosmology says that during this epoch the forces that were once contained inside the singularity point begin to seperate.
Gravity was thought to have been the first to seperate. While the Nuclear strong force, Nuclear weak force and Electro-Magnetic forces were still united. This unification of the 3 forces is known as the GUT force.

Cosmology's equations for this epoch are based upon the energy & properties of a Photon. However it seems that at this early point in the development of the universe the Photon was not yet formed. This means that its energy was not available to be utilised at this point. If a Photon was not present at this time, something else must have been present. The Nucleon-Deuteron modle finds that at this stage the Majorana singularity particle should have been present, the temperature should have been as hot as 10^{32} degrees kelvin rapidly cooling to 10^{27} degrees kelvin.

Space time as we know it is still yet to form but the Majorana singularity particle should have the ability to spin as particles do, creating its own unique anapole energy. The speed at which it spins should be 2.9979×10^{-8} m/sec which is close to the speed of light in a vacuum.

New forces are created when the Majorana singularity point experiences its heat sink process and stretches out into a 'do-nut' shaped stucture. To begin with the centremost point should become elongated. The centre stays as a neutral area, with zero properties similar to the centre of a black hole. This

has properties similar to the Z Boson that we find inside of Atoms.
The Majorana singularity particle has a wavelength of 1.6000X10^-35mm, it stretches to 1.6000X10^-30mm by the end of the epoch.
To find the wavelength λ for the end of this era:
$$\lambda = \frac{0.0029}{T\ 1.8125 \times 10^{27} k}$$
wavelength λ = 1.6000x10^-30mm

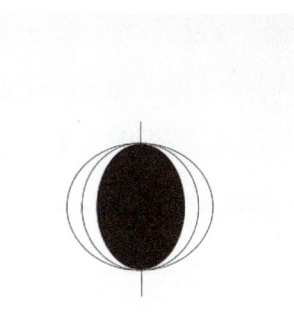

Diagram showing Majoranas anapole protective membrane.

The Majorana is taking on both particle and wave properties as it's state shifts into an elongated 'do-nut' shape, it is forming into a Dipole structure. The outer most parts at the edges begin to develop primitive poler reigions. The energy flow originating and traveling from the area of higher temperature to the lower area. This process must create the Anapole state that is observed in the Majorana particle. This should have eventualy led to the creation of the Electro-weak force and should be the cause of the next Electro-weak era.

The Majorana heatsink process could be thought of as the temperatutre change from Plank to GUT era. 10^{-32} kelvin to 10^{-27} kelvin. While the temperature is dropping the central vacuum area of the Majorana particle is being stretched out to reach critical stretching point.

Diagram showing Majorana particle gravitational type force.

At this point the particle may have formed a protective membrane around itself to help keep the structure together. This would be the first evidence of particle Self preservation. Over stretching may have caused the vacuum to create poles that will attract to each other as a way to keep the structure together as the structure becomes dipole. Cosmology traditionaly says that Gravity seperated from the Superforce during this epoch. The Nucleon-Deuteron model suggests that during this epoch the Expanding Majorana singularity point might have experienced some sort of gravitational pull within its own structure. This could have been its attempt to hold itself together to keep itself from being torn appart by the opposing temperature fluctuation caused by heat sink.

Gravity would not have existed outside of this structure because the structure itself was all that existed. So at this point gravity as we know it would not have existed as a force which draws two masses together as we see in stars and planets.

29
Majorana V Black Hole

Results of Research into the milky way black hole from observations made by Hubble find that Black holes have a Dark matter ring around the outer edge of the black hole structure. This is thought to act as a membrane which holds the structure together. Like wise the Majorana particles may also have a protective membrane around its perimeter in its 'do-nut' shape, this is because the 'do-nut' structure is forming a natural dipole. We know that dipoles cause Electro-Magnetic forces to form. The presence of such a force around the dipole shape would create a natural energy membrane around it similar to the structure we see in celestial Black Holes.

Black Holes are neutral at their core which are negative, with zero value. The singularity particle was thought to have started out as a neutral state, of zero value that then stretched or inflated. Black Holes stretch out and become larger. When a black hole is at the center of a galaxy the whole structure is often observed to have a 'do-nut' shape, galaxies also tend to have a double lobe structure, which is similer to the fully stretched out or dipoled Majorana form. Super-nova explosions reveal a similar structure with a black hole forming in the mid section. (as seen on front cover)

The core of a black hole is where matter is reduced to or reaches 0 point and is surrounded by a halo of light which is the event horizon. The event horizon seems to be caused by the matter becoming reduced at impact on contact with the 0 vaccum center zone. The matter that is pulled toward the black hole has been observed by hubble as being a spiraling mass. The impact causes gamma rays to be produced and Neutrinos are released. In a similar way as the Majorana singularity particle should produce Neutrinos. However the obvious difference is that the Black Hole produces many more at any one time.

In the Nucleon-Deuteron modle the Majorana is depicted as having a neutral center, zero state, which is a micro void or micro Black Hole. This Micro Black Hole is depicted as being surrounded by a very elongated transition zone or primitive "event horizon". It shall become the Z Boson. Due to stretching the middle of the 'do-nut' (micro black hole) may naturally become compressed forming flux tube like structures which reach from the centre to each dipole end. These are the points of maximum critical stretching.
The edges of the transition zone seem to have developed into polar ends. One is positve the other is negative.
In the model the outer edge of the Majorana particle is depicted as having a protective membrane similar to that of a Black Hole.

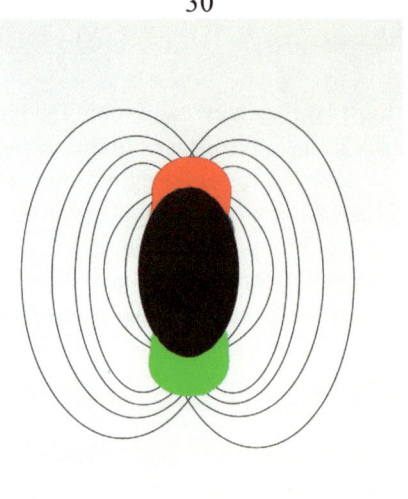

Diagram showing Majorana as Anapole state becoming Dipole.

The Majorana Singularity V Sacred Geometry

Sacred Geometry is an ancient art. It was developed in ancient mystery schools. The ancients seemed to have had access to knowledge that has become lost to us over time. Some of this ancient knowledge is now being re-discovered and re-understood. Sacred Geometry says that there are 5 naturally occurring solid platonic 3D shapes which make up all things in the universe. These shapes are:
Tetrahedron
Hexahedron
Octahedron
Dodecahedron
Icosahedron.

Sacred Geometry says that the universe began with a single point. Eastern religions call this point the cosmic egg. This can be very losley compared with the Majorana particle.

Sacred Geometry says that the point was a single point of primitive consciousness which couldn't experience anything because it had no point of reference in a void. So the point created a circle around itself, almost like a membrane.

This sounds similar to the idea that the Majorana formed a protective membrane around itself during the heat sink process where it began stretching out. Sacred geometry says that the point moves to the edge of the circle then creates another circle. This is a Vesica Piscis.

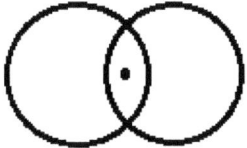

Ancient civilizations also considered the vesica Piscis as representing the joining of God & Goddess to create their offspring.
In ancient Buddhism it was thought that the universe began with Yin & Yang. These are a male and female energies that are equal to each other. The two energies flow around each other forever entangled. Each half holds within itself the seed of the other.
The Nucleon- Deuteron model shows that there are two such energies which are entangled with each other. Although we might consider them as energies to our modern way of thinking, rather than as Gods.

The Majorana particle could be compared to a vesica piscis which elongated to become a 'do-nut' shaped Vesica piscis. This is because when the particle stretched it must have formed polar regions at each end.

These poles become the Head and Tail ends of Neutrino and Anti-Neutrino entangled pairs. The polar energy that the Head (+1) and (-1) tail produce, causes a Vesica piscis shape around the Neutrino pair.

This energy stretches out to incorporate the next (+) head or (-) tail in the Quark structure. This seems to be what causes the Quarks shape.

Sacred Geometry says that the point keeps traveling to the edge of each new circle, only to create the next circle that it can travel into. The final structure is known as the flower of life.

The Ancients might have had the right idea. The Nucleon-Deuteron model shows that the energy of the Neutrinos, moves from one circular head or tail to the next. As it does this a structure is formed. However in the case of the Nucleon-Deuteron model the structure is not a flower but a spiraling structure which eventually creates a cube-like structure of the Nucleon or Deuteron.

This tiny Majorana singularity particle does not seem to annihilate as previous science theories have predicted. The process happens so fast that it is difficult to detect what happens to the Majorana singularity particle. If we keep this in mind together with the laws of Thermodynamics, we can see that the Majorana singularity particle does not seem to annihilate but changes form as a result of the heat sink process.

The Majorana particle never seems to make a complete split into two separate particles. Upon stretching to critical point its poles become further apart and its center elongated and stretched into a long gluon type flux tubular structure which keeps the polar ends together.

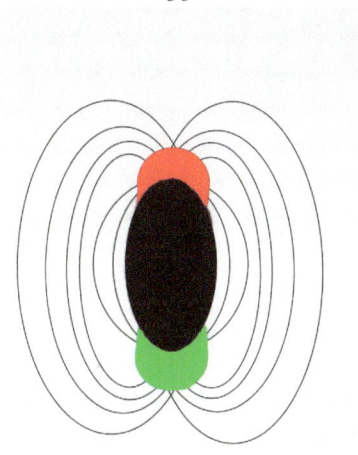

Diagram showing Majorana anpole structure becoming dipole with spiraling w+ & W- Bosons.

The area of the gluon type flux tube near the Neutrino head end will have (+) polarity. This will become a W+ Boson. Energy flows along these gluon/ Boson flux tubes. The centre most point always remains as a vacuum area with 0 properties and will become a Z Boson. Its nature is such that Electro-Magnetic energy can travel from the Majorana Neutrino positive dipole end thorough the Z Boson into the Majorana Neutrino negative dipole end during the Electro-weak era. Creating an Electro-Magnetic field around the particle. We know that when a particle decays/changes state energy is always conserved this results in the emission of Neutrinos. The Majorana singularity particle should be no exeption and should also emmit Neutrinos when it decays/changes state. It is now known that Neutrinos do carry a very small mass.

The Majorana singularity point has expanded / Stretched from Plank length of 10^{-35}m to 1.6000×10^{-30}mm. This should be the size of the Majorana singularity particle at its 'do-nut' stage. These lengths are interchangable with wavelength at this point because the universe is no bigger than these lengths and so a wave could not be any larger than this size.

As we know traditionaly science uses the Photon to describe these early conditions, however the model shows that a Majorana singularity point and its Neutrino have very similar properties to a Photon. Therefore we can assume that the Majorana singularity partcle and its Neutrino pairs can exist in both particle & waveform and have wavelengths which can be no bigger than the size of the universe it is in. The whole universe is now only as big as the Majorana particles wavelength.

Summery Of The Equations For GUT Era/Epoch

Equations for GUT epoch:
C= speed of light 2.99792458×10^8 (metres per sec)
MSS = Majorana singularity speed/ velocity (metres per sec) 2.9979×10^8 m/sec
MNPS= Majorana Neurtino pair speed 2.9979×10^{-8} m/sec
$\sqrt{Gh}$= plank time $<1.35 \times 10^{-43}$(seconds)
pl = plank length 1.6000×10^{-35}(mm)
h= planks constant
T= temperature (Degrees Kelvin)
t= time (seconds)
λ= wavelength (mm)
f= frequencey (HZ)
E= energy (Joules or Gigga electron volts)
v= velocity (metres per sec)

We know the temperature at GUT epoch to be 10^{27}k, we can use this to chart the wavelength of the Gut epoch and the Majorana particle throughout its temperature change until the end of Gut epoch: Gut epoch began at 1.8125×10^{32}k -ended at 1.8125×10^{27}k.

At 1.8125×10^{32}k the wavelength was 1.6000×10^{-35}mm
At 1.8125×10^{31}k the wavelength was 1.6000×10^{-34}mm
At 1.8125×10^{30}k the wavelength was 1.6000×10^{-33}mm
At 1.8125×10^{29}k the wavelength was 1.6000×10^{-32}mm
At 1.8125×10^{28}k the wavelength was 1.6000×10^{-31}mm
At 1.8125×10^{27}k the wavelength was 1.6000×10^{-30}mm

we can see how the wavelength changed during this epoch.
As the wavelength changed so did the structure of the particle.

To find the frequencey at the end of Gut epoch:
$$F= \frac{C}{\lambda\ 1.6000 \times 10^{-30}mm}$$
Frequencey = 1.8737×10^{38}HZ

Now we have the frequencey of the Majorana particle at the end of the epoch, we can work out its velocity by multiplying its wavelength by its frequencey.
V= $\lambda\ 1.6000 \times 10^{-30}$mm x f$1.8737 \times 10^{38}$HZ
velocity = 2.9979×10^8 m/per/sec

To find the Energy at the end of the Gut Epoch & the Majorana singulaity particle:

$$\frac{h \times MSS\ 2.9979 \times 10^8}{\lambda 1.6000 \times 10^{-30} mm}$$
Energy = $1.2415 \times 10^5 J$

or

E = h × f 1.8737×10^{38} HZ
Energy = $1.2415 \times 10^5 J$

Traditionaly the equation using light speed would be:

Mass = $\frac{Energy\ 1.2415 \times 10^5 J}{light\ speed(c^2)\ 3 \times 10^{8^2} m/sec}$
= Mass $1.3814 \times 10^{-12} kg$

To find the mass of the Majorana we should use its own speed.

The formula is: Mass (m) = $\frac{Energy\ (E)}{Majorana\ speed^2\ (msp^2)}$

so

Majorana Mass = $\frac{Majorana\ Energy\ 1.2415 \times 10^5 J}{MSS\ 2.9979 \times 10^{8^2}\ m/sec}$
= Majorana Mass: $1.3814 \times 10^{-12} kg$

Because E = MC²

E = $1.3814 \times 10^{-12} kg \times 2.9979 \times 10^{8^2}\ m/sec$
E = $1.2415 \times 10^5 J$

36
Initial Inflation Period

Cosmology's standard Model says that there was one inflation epoch at this time. The nucleon-Deuteron Model finds that there seems to be an initial inflationary Epoch then another a little later.

During this Period of Initial inflation, it is thought that all Mass and Energy was created. This Period began at 10^{-35} Seconds and lasted until 10^{-32} Seconds. This Epoch happened simultaiously with the Electro-weak epoch but was shorter lived and ended before the Electro-weak era ended. The Temperature at this time was 10^{27} degrees Kelvin.

The equations for Electro-weak era also run concurrent with this Epoch.

The Cosmic Neutrino Background radiation is thought to detect Neutrino decoupling. The event happened within the first second of the universe lifespan. It is thought to have been a consequence of weak interactions between Neutrinos. This could have happened durng the time of the initial inflation period where Primordial Majorana Neutrinos were reproducing at a fast rate.

The Nucleon-Deuteron Model finds the following equations to describe the Initial Inflation epoch:

To find the wavelength at the beginnging of the Initial Inflation period:
$$\lambda = \frac{0.0029}{T1.8125 \times 10^{26} k}$$
$$\lambda = 1.6000 \times 10^{-29} m$$

To find the Frequencey at the beginning of the Initial inflation period:
We traditionaly use the equation light speed (C) over wavelength (λ).
However as we know the Photon had not been formed at this time. The Nucleon-Deuteron model suggests that at this point, Majorana singularity was stretching with an increased wavelength of $1.6000 \times 10^{-29} m$. This is longer than the wavelength of the Majorana singulaity but still too short to be a Neutrino which are $1.6000 \times 10^{-24} m$. This is an in between stage where the particle is in late 'do-nut' stage. It is now almost a fully formed dipole particle.

How The Majorana 'Do-nut' Caused The 'Initial' Inflation Period.

Cosmology finds that our universe has the same wavelength in all directions. In order for the CMB of our universe to be in this state cosmology says that it must have grown extremely fast in a uniform way at a very early stage in the universe. The universe was said to have inflated by a factor of 10^{50}. This takes it to the size of a galaxy. This also means that the universe seems to have inflated faster than the speed of light. This causes a problem because nothing is supposed to travel faster than light speed. The Nucleon-Deuteron model finds how and why this happened. Inflation must have taken place, however it might have happened in two stages. This is stage one the initial Inflation period. Here the universe first experiences a notable change in size.

The secondary Inflation period is where the universe inflated to the factor of 10^{50}. This major event seems to have taken place a little later on. The Nucleon-Deuteron model finds that at this point where Majorana particles are becoming dipole due to the heat sink process, some energy is being lost or emited. This energy is emited in the form of more particles. These particles should be emited in all directions.

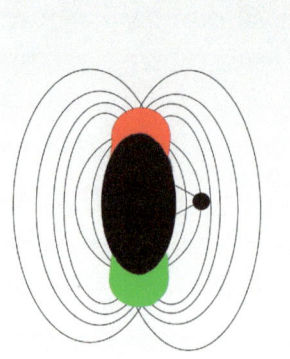

Majorana 'dou-nut' anapole particle emiting energy in the form of a new particle.

The diagrams show the emission of only one particle to make the diagrams clearer. Because as the diagrams progress they become busier. Therfore to many particles being emitted on them would result in unclear diagrams. We shall imagine that this process is happening with the particle emitions numbering aprox 10,000 particles. We know this because if the Majorana particle displays similar charcteristics to the Photon, it must contain enough energy to produce at least 10,000 particles.

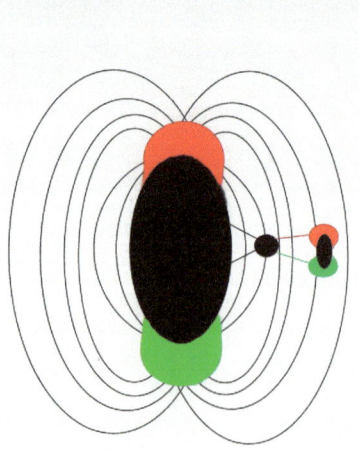

Majorana 'dou-nut' particle
emitting new particles which becoms anapole.

The Nucleon-Deuteron model shows how the original Primordial Father /Primordial Mother Majorana Neutrino in its new state "births"/emits another particle generation which will grow to become more Neutrino/ Anti-Neutrino pairs. The new pairs might be emitted from the Z Boson 0 vaccum point. This may happen because stretching inside the Neutrino body causes some energy
from head (+) 1 & tail (-) 1 to be forced into the mid section creating a leak or burst.

This emitted energy would naturally be smaller than the Majorana particles present late 'do-nut' shape. It should be as small or almost as small the Majorana singularity particle. This emitted particle should then be capable of experiencing the same change of state as its parent Majorana particle did. It should go through the anapole stage then rapidly become dipole. This process should take 1.0000×10^{-8} sec. Because this is how long it takes for the origional Majorana particle to decay from its initial state to 'do-nut' formation.

These particles are to become dipole Neutrino pairs which should cause Electro-Weak force. The primordial Father & Mother continue to grow into a full size Majorana Neutrino entangled pair.

By the end of the initial Inflation period the universe has thousands of new particles in it and is much larger for this reason. The new particles should have been emitted and propelled at a velocity of 2.9979×10^8 m/sec, from the spinning Majorana particle that was changing from singularity state through anapole to dipole state, via the heat sink process of Thermodynamics.

Therefore the size of the universe at the end of the initial Inflation period should have been 2.9979×10^5 metres. Because Velocity 2.9979×10^8 m/sec x Time 1.0000×10^{-3} seconds = a distence of 2.9979×10^5 metres. This size is 10,000 metres.

As we know the first Inflation period began at 10^{-35}
seconds and lasted until 10^{-32} seconds.
The universe seems to have inflated due to Majorana Neutrino pair replication. Although each individual particle did not travel faster than 2.9979×10^8 metres per second, each particle should have projected copies of itself out at the velocity of 2.9979×10^8 metres per second.

To an imaginary observer the initial inflation event should appear to be unfolding faster than the speed of light.

The first Primordial Father & Mother Majorana particles should have each contained enough energy to reproduce at least 10,000 more copies of itself.

These particles would have been emitted equally in all directions due to the spinning motion of the primordial Father & Mother entangled pair. Each copy being propelled at a velocity of 2.9979×10^8 m/sec should have been capable of reproducing 10,000 more and so on. With hundreds of thousands of particles all doing this, we can see how the universe inflated to 10,000 metres so fast.

The process of replication should have continued until the rate of replication over took the particles escape velocity. This would mean that the new particles were being replicated faster than they could repel each other and could no longer escape each others grip. This should have resulted in the particles becoming more densely packed into the space that they are creating. This would result in the Neutrinos being forced to flip in flight. The flipping process would naturally cause the pairs to attract, thus creating Electro-Magnetism.

When the initial Inflation process slowed due to density the overall velocity of each particle was still 2.9979×10^8 metres per second.

Summery Of The Equations For The Initial Inflation Period :

Equations for the initial inflation period :
C= speed of light 2.99792458x10^8 (metres per sec),
MSS = Majorana singularity speed/ velocity (metres per sec) 2.9979x10^8 m/sec
MNPS= Majorana Neutrino pair speed 2.9979x10^8 m/sec
√Gh= plank time <1.35x10^-43(seconds)
pl = plank length 1.6000x10^-35(mm)
h= planks constant
T= temperature (Degrees Kelvin)
t= time (seconds)
λ= wavelength (mm)
f= frequencey (HZ)
E= energy (Joules or Gigga electron volts)
v= velocity (metres per sec)

We know that Electro-Weak epoch began at around the same time as this Inflation period. Electro-Weak epoch began at a temperature of 10^27k. The initial Inflation period should have started just after. This should give the initial Inflation period a temperature of 10^26k.

If the temperature at the begining of the Initial Inflation was 10^26k, it should have dropped to 10^21k by the end.

we can use this to chart the wavelength of the period and the Majorana particle throughout its temperature change until the end of the Initial inflation period:

At 1.8125x10^26k the wavelength was 1.6000x10^-29mm
At 1.8125x10^25k the wavelength was 1.6000x10^-28mm
At 1.8125x10^24k the wavelength was 1.6000x10^-27mm
At 1.8125x10^23k the wavelength was 1.6000x10^-26mm
At 1.8125x10^22k the wavelength was 1.6000x10^-25mm

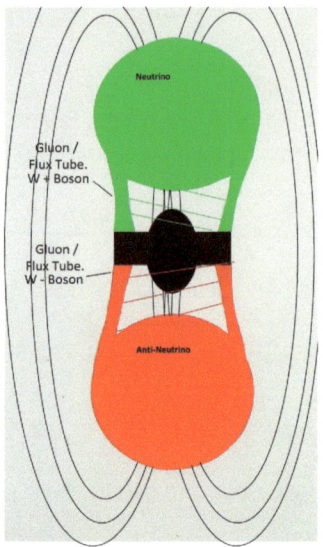

Fully dipole Neutrino/Anti-Neutrino pair.

At 1.8125×10^{21}k the wavelength was 1.6000×10^{-24}mm (yoctometre) we can see how the wavelength changed during this epoch. As the wavelength changed so did the structure of the particle. The Majorana Do-nut shaped particle changed its wavelength from 1.6000×10^{-29}mm to 1.6000×10^{-24}mm which is a yoctometre. We know that the yocktometre is the size of a Neutrino.

The frequencey at the begining of this epoch would have been 1.8737×10^{37}HZ.

To find the frequencey at the end of the Initial Inflationary period:
we would traditionaly use the equation light speed (C) over wavelength (λ). However the Nucleon-Deuteron model uses the speed of Majorana Neutrino pairs (MNPS) the equation is as follows:

$$\frac{(MNPS) 2.9979 \times 10^{8}}{\lambda\ 1.6000 \times 10^{-24}m} = \text{Yoctometre}$$
$$= \text{Frequencey } 1.8737 \times 10^{32}\text{HZ}$$

We know the frequencey of the Majorana particle at the end of the initial Inflation period to be 1.8737×10^{32}HZ so we can work out its velocity by multiplying its wavelength by its frequencey.

To find the velocity of the Majorana in its 'do-nut' state during the Initial Inflation period:

We have to take the wavelength at this time 1.6000×10^{-29}m and multiply it by the frequency at this time which was 1.8737×10^{37}HZ.
So
1.6000×10^{-29}m x 1.8737×10^{37}HZ = 2.9979×10^{8} m/sec.

This means that the universe during inflation, expanded with a velocity which began at 2.9979×10^{8}m/sec.
The conditions changed during this period. To calculate the velocity at the end of the period we first need to find the wavelength for this time.

For this we need to use Wiens law:
$$\text{Velocity} = \frac{0.0029}{\text{Temperature } 1.8125 \times 10^{21} \text{ Kelvin}}$$
= 1.6000×10^{-24}m = yocktometre = the size of a Neutrino.

V= λ 1.6000×10^{-24}mm x f1.8737×10^{32}HZ
velocity = 2.9979×10^{8} m/per/sec

We can see that the velocity of the Majorana singularity particle remained constant throughout its changes. It maintaines the same velocity in its new Neutrino pair, dipole state.

To find the Energy at the end of the period & the Majorana Neutrino pair we can now use the velocity or Majorana Neutrino pair speed (MNPS) because that is what now exists:

E= h x MNPS $\frac{2.9979 \times 10^{8}}{\lambda \ 1.6000 \times 10^{-24}\text{mm}}$
Energy = 1.2415×10^{-1}J
or
E= x f1.8737×10^{32}HZ
Energy = 1.2415×10^{-1}J

To find the mass of the Majorana Neutrino pair we should use their own speed.
Majorana Mass = Majorana Energy $\frac{1.2415 \times 10^{1}\text{J}}{\text{MSS } 2.9979 \times 10^{8^{2}} \text{ m/sec}}$
= Majorana Mass: 1.3814×10^{-16}kg
Because E=MC²
E= 1.3814×10^{-16}kg x $2.9979 \times 10^{8^{2}}$ m/sec
= 1.2415×10^{1}J

To find the velocity at the end of the Initial Inflation period: we need to multiply the wavelength by the frequencey:

$$\lambda\ 1.6000 \times 10^{-24}m \times f\ 1.8737 \times 10^{32} Hz$$
$$= \text{Velocity of } 2.9979 \times 10^{8}\ m/sec$$

To find the Distence covered & size of the universe at the end of the epoch: we could multiply the velocity at the end of the epoch by time:

$$\text{velocity of } 2.9979 \times 10^{8} m/sec \times \text{Time } 1.0000 \times 10^{-3}\ sec$$
$$= \text{Distence of } 2.9979 \times 10^{5}\ \text{metres. Which is equil to } 10,000km.$$

To check the time we divide distence by velocity:
$$\frac{D\ 2.9979 \times 10^{5}\ \text{metres}}{V\ 2.9979 \times 10^{8} m/sec}$$
$$\text{Time} = 1.0000 \times 10^{-3}\ sec$$

Another way to find the velocity at the end of the epoch is to divide distence by time:
$$\frac{D\ 2.9979 \times 10^{5}\ \text{metres}}{\text{Time} = 1.0000 \times 10^{-3}\ sec}$$
$$\text{Velocity} = 2.9979 \times 10^{8} m/sec$$

The end of this epoch should have taken 10^{-3} seconds for inflation to slow back down.

If the time at the begining of the initial Inflation period was 10^{-35} seconds and lasted until 10^{-32} seconds, the period would have lasted for 10^{-3} seconds.

So the Initial Inflation period began at the length of $10^{-30}m$ and inflated to 2.9979×10^{5} metres which is equil to 10,000km. This happened at a time before Photons existed and so no light / radiation from this period should be detected in the cosmic microwave background radiation signal. The inflation period that is seen in the CMB would have been produced in the later secondary inflation period that should have taken place when Photons became free during the Decoupling Epoch. This is possible because all energy distribution should be equally distributed until then.

At the end of the era when the universe stopped the accelaration process due to density, the velocity was 2.9979×10^{8} m/sec. This is the speed of Majorana Neutrino pairs.

Neutrino Flip And Baryogenisis
In Electro-Weak Era/Epoch

The discovery that Neutrinos flip was discovered by Takaaki Kajita & Arthur McDonald at underground facilities in Japan and Canada. They were awarded the 2015 Nobel Prize for Physics for their discovery. They theorised that Neutrinos flip, and that this action causes oscillations during "flight".

The Nucleon-Deuteron model suggests that as the temperature cools, conditions are becoming more dense.
Particles are being emitted faster than the existing particles can repel from each other. This causes a density. The particles find themselves tightly packed and so repulsion is no longer possible.
The law of attraction causes the Neutrino pairs to flip in flight so that positive & negative become attracted to each other. This attraction is the process which causes Electro-Magnetism and Quarks to be produced. This is the beginning of Matter in the early universe. The conditions have been described as a cooling Quark gluon plasma.

Neutrino pairs flip & attract.

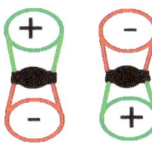

The Heisenberg uncertainty principle has traditionaly been used to describe the process of matter formation & shows the amount of time that a partcle and its anti-particle might be in existence before they annihilate each other. This principle was based on the assumption that these particles & Anti-particles fizz in and out of existence via annihilation. Some of them stick and some do not which in this principle, it is thought that this process might lead to there being more Anti-matter than normal matter in the universe. The principle uses virtual particles which are theorised particles that lay beyond our ability to detect. They have since been identified as Anti-particles.

However the Nucleon-Deuteron model suggests that particles & anti-particles might not completley fizz out of existence because energy must be conserved. Therefore if a particle and anti particle annihilated each other, they in theory must conserve their energy in another form.

Science has found in numerous experiments that when particles seem to annihilate they produce Neutrinos as a way of conserving their energy.

The Heisenberg uncertainty principle is important because it shows how long the Majorana Neutrino and its Anti Neutrino must take before they change state. Rather than annihilating each other as Heisenberg predicts. The particles seem to replicate to conserve their energy. Therfore Heisenberg annihilation/uncertainty time can be thought of as replication time for the purpose of the Nucleon-Deuteron model.

Heisenberg uncertainty principle says that particles and Anti particles must have annihilated each other before 35×10^{-25} seconds. This is known as the uncertainty of time Δt.

Science says that radiation should be effected by Inflation. The effect should be polerisation. Evidence for this was found in 2014. A signal was detected in the CMB which showed that at this time of inflation polerisation did happen.

Electro-Weak Era/Epoch

Electro-Weak epoch happened simultanously with the initial Inflation period. Inflation period ended at 10^32 seconds while Electro-weak era continued until 10^-12 seconds. Electro-Weak epoch began at a temperature of 10^27k and ended at a temperature of 15k.

During the beginning of the Electro-Weak epoch more Majorana Neutrinos are being emitted.

The Neutrino's are being repelled from each other. By the end of the epoch the Neutrino's might find that there are so many of them in such a small

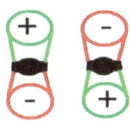

volume of space that they can no longer repel but have to Flip.
To find the frequency the equatuion is as follows:

$$f = \frac{MNPSP\ 2.9979 \times 10^{8} m/sec}{\lambda\ 1.6000 \times 10{-29} mm}$$
$$= f\ 1.8737 \times 10^{37} HZ$$

So at this time the frequencey was $1.8737 \times 10^{37} HZ$.

During the Electro-Weak epoch, the Electro-Magnetic force seems to have been created rather than separating from the Super-force. The Neutrinos spew out new pairs into the local vicinity causing space time to expand. Soon the numbers of pairs quickly mount up. Each pair are travelling at close to light speed. When Dipole Neutrino pairs emit new pairs, the rate of replication is faster than the velocity of the expanding universe. This is because the new pairs will be projected at a velocity of close to light speed as well as travelling themselves at close to light speed. The result of many of these particles being formed and projected in this way should create a faster than light effect. The universe seems to have expanded faster than light speed. This might have happened in a steady, even stream. While early Majorana Neutrinos were becoming more densely populated they flipped and formed into Quarks. Their off spring travelled out a little more until their region became dense at which point they too formed Quarks. The universe continued to grow, but at a slower rate. This might have been due to its density & the Electro-Magnetic force capturing Neutrino pairs to form Quarks.

The Neutrinos would flip in flight so that the positive of one set would be in contact with the negative of another set. This would result in attraction and an Electro-magnetic force would be created. The Electro-Weak era is still happening and ends at 10^{-12} Seconds at a temperature of 10^{15} Degrees Kelvin. The Temperature would also be dropping during this process because as the rate of inflation is slowing down. This action would create a spiral chain/string like effect which would result in Quark formation. This might be why Quarks seem to be comprised of strings. The new force which bound the Neutrinos and held the Quarks together would be known as Nuclear Weak Force.

So here we could say that Nuclear Weak force was created rather than being seperated from the superforce.

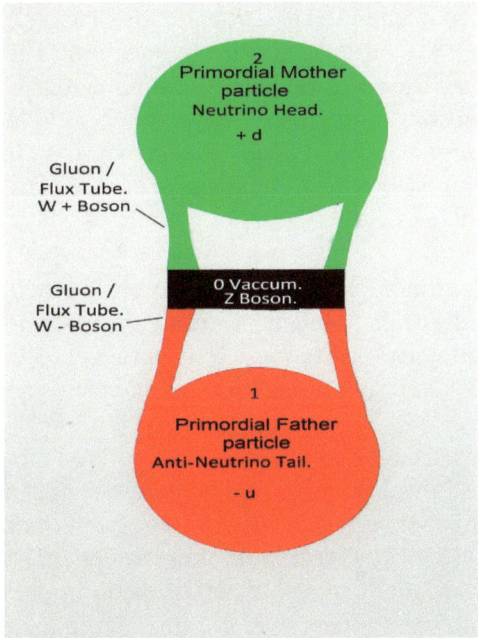

The first Neutrino/Anti-Neutrino pair are born.

Neutrino Pair Attributes

Both Neutrino and Anti-Neutrino would each have one half of unity because they are still technically one united particle which exist at different states at different "locations" of its body.

The Neutrino head end of the entangled pair is positive with the value (+1) and 1/2 spin of unity. The Neutrino is usually thought of as being "Left handed" because of the direction of its half spin of unity. For this Model this "left handedness" will be depicted as an up (+1) Head spinning to the left.

The positive Neutrino head end is thought of as neutral and dormant until it interacts with a negative Anti-Neutrino counterpart.

The Anti-Neutrino tail end of the Neutrino entangled pair is negative with the value (-1) and spin 1/2 of unity. The Anti-Neutrino is thought of as being "Right handed" because of its half spin direction of unity. For this model "Right-handedness" will be depicted as a down (-1) tail spinning to the Right. It carries most of the information and therefore most of the mass of the pair. It's interaction with the positive Neutrino head causes the positive head to become active.

The poles (+) & (-) have been stretched to maximum point. The middle section still held 0 properties and remained a vacuum zone, the Z Boson.

This dipole structure now has a Neutrino positive end which experiences its own angular momentum. The structure also has an Anti-Neutrino Negative end wich also has its own angular momentum. The two working together through the Z Boson gives the Neutrino entangled pair their magnetic moment. Z Bosons must exists in large numbers in large atoms. When atoms are smashed together in a collider, the gravitationl effect which the Z bosons should have on each other should cause them to gravitate together possibly seeming as one particle.

When the Neutrinos flip and attract to each other Magnetism is created. Electro-weak force & Magnetism results in Electro-Magnetism. And so the first Electro-Magnetic wave/particle is born as the first Quark. Because two Neutrino/ Anti- Neutrino bound pairs should create a Quark.

During the construction of the Nucleon-Deuteron model it was noted that the interactions of the Neutrinos seem to mirror the natural world. We find similar patterns between genders. The eggs of a female usually lie dormant inside a womb. They also hold information. The Sperm is the active counterpart who brings the other half of the information and the spark which ignites life into the dormant egg. For this reason the positive Neutrino heads which are dormant until they interact with Anti-Neutrino tails, have been depicted as female while the negative Anti-Neutrino tail ends have been depicted as male. This contradicts scientific and eastern cultural thought where male energy is positive and female energy is negative.

Naming Neutrino Generations

In order to differentiate between evolving primordial Neutrino generations, each positive Neutrino head & negative tail have been given nicknames.

The first generation have been given the names (-1) Primordial Father & (+1) Primordial Mother Neutrino pair. This is because they seem to be the first particles to form after the Initial vacuum Singularity state of the early universe. They were the first particle that had an intent to survive and had the ability to replicate. Their programming seems to have influenced and permeated every atom in the universe. Although these names might not be to everyone's taste, they are easily memorable.

Neutrino pairs in their Generations in Hadron & Quark families.

Belonging to the Neutron:

Generation 1: Quark family 1 = Primordial Father & Primordial Mother
Generation 2: Quark family 1 -2= Adam & Evelyn
Generation 3: Quark family 2 = Cain & Claire
Generation 4: Quark family 3 = Able & Abby
Generation 5: Quark family 3-4 = Brian & (Beth)

Belonging to the Proton: (inc Beth)

Generation 6: Quark family 4 = Charles & Charlie
Generation 7: Quark family 5 = Dan & Diana
Generation 8: Quark family 5-6 = Eddie & Edna
Generation 9: Quark family 6 = Freddie & Frayer
Generation 10:Quark meson family 7 = George & Georgina.

Summery Of The Equations For The Electro-Weak Era/Epoch

Equations for the Electro-Weak era:
C= speed of light 2.99792458x10^8 (metres per sec)
MSS = Majorana singularity speed/ velocity (metres per sec) 2.9979x10^8
MNPS= Majorana Neurtino pair speed 2.9979x10^8
$\sqrt{Gh}$= plank time <1.35x10^-43(seconds)
pl = plank length 1.6000x10^-35(mm)
h= planks constant
T= temperature (Degrees Kelvin)
t= time (seconds)
λ= wavelength (mm)
f= frequencey (HZ)
E= energy (Joules or Gigga electron volts)
v= velocity (metres per sec)

We know that Electro-Weak epoch happened simultanously with the initial Inflation period. Inflation period ended at 10^32 seconds while Electro-weak era continued until 10^-12 seconds. Electro-Weak epoch began at a temperature of 10^27k and ended at a temperature of 15k.

We can use this to chart the wavelength of the period and the Majorana particle throughout its temperature change until the end of the initial Inflationary period:

At 1.8125x10^27k the wavelength was 1.6000x10^-30mm
At 1.8125x10^26k the wavelength was 1.6000x10^-29mm
At 1.8125x10^25k the wavelength was 1.6000x10^-28mm
At 1.8125x10^24k the wavelength was 1.6000x10^-27mm
At 1.8125x10^23k the wavelength was 1.6000x10^-26mm
At 1.8125x10^22k the wavelength was 1.6000x10^-25mm
At 1.8125x10^21k the wavelength was 1.6000x10^-24mm (yoctometre) size of a Neutrino.
At 1.8125x10^20k the wavelength was 1.6000x10^-23mm
At 1.8125x10^19k the wavelength was 1.6000x10^-22mm
At 1.8125x10^18k the wavelength was 1.6000x10^-21mm
At 1.8125x10^17k the wavelength was 1.6000x10^-20mm
At 1.8125x10^16k the wavelength was 1.6000x10^-19mm
At 1.8125x10^15k the wavelength was 1.6000x10^-18mm

To find the frequencey at the end of the Electro-Weak era:

$$F = \frac{MSS}{\lambda} = \frac{2.9979 \times 10^8}{1.6000 \times 10^{-18} mm}$$

Frequencey = 1.8737x10^26HZ

Now we have the frequencey of the Majorana Neutrino entangled dipole pair. From this we can work out its velocity by multiplying its wavelength by its frequencey.

V= λ 1.6000×10^{-18}mm x f1.8737×10^{26}HZ
  velocity = 2.9979×10^8 m/per/sec

Again this shows that the velocity of the Majorana singularity particle remained constant throughout its changes. It maintaines the same velocity in its new Neutrino pair, dipole state.

To find the Energy at the end of the period & the Majorana Neutrino pair we can now use the velocity or Majorana Neutrino pair speed (MNPS) because that is what now exists:

E= h x MNPS $\underline{2.9979 \times 10^8}$
  λ 1.6000×10^{-18}mm
Energy = 1.2415×10^{-7}J

or
E= h x f1.8737×10^{26}HZ
Energy = 1.2415×10^{-7}J

To find the mass of the Majorana Neutrino pair we should use their own speed.

Neutrino pair = Majorana Energy $\underline{1.2415 \times 10^{-7}J}$
                  MSS $2.9979 \times 10^{8^2}$ m/sec
Neutrino pair mass = 1.3814×10^{-24}kg

Because E=MC²
  E= 1.3814×10^{-24}kg x $2.9979 \times 10^{8^2}$ m/sec
  E= 1.2415×10^{-7}J

Energy per mole = Energy x Avagadros constant (NA)
so
Energy per mole = 1.2415×10^{-7}J x 7.4765×10^{16}J
  E = 9.2821×10^9J

Bosons & Flux-Tubes V Time Passage & Beta Minus Decay

Little is known about Bosons, however it is thought that the mixing angle where spontaneous symmetry is thought to rotate produces the Z Boson. This mixing angle seems to be the change of state that occures in the gluon flux tube which turns a W+ Boson over to become a W- Boson though the vacuum reigion, Z Boson. This process seems to give rise to flavors of both Neutrinos and Quarks. It should be the cause of the first osillations in the early universe.

The tiny Gluon/ Flux tubes join the positive Neutrino heads & negative Anti-Neutrino tails together. When one pair of Neutrinos emit another pair the two could become attracted together to form Quarks. The W- & W+ Bosons seem to be the cause of nuclear Weak force & Entanglement. They cause Beta minus decay, the production of stong nuclear force and play a fundamental role in the Nucleon's replication system. They also help to give the Quarks and ultimatlly the final atom mass.

The flux tube / gluon / W- & W+ Bosons will be dipicted as spinning tubular bodies. The poitive Neutrino head ends are green and their positive spining energy states along the flux tube/ W+ Boson are depicted as green. The Anti-Neutrino tails are depicted as red and their negative spining energy states are dipicted as red along the gluons/flux tubes/ W- Bosons. The W- & W+ Bosons/flux tubes/ gluons act as transition zones between energy states, they have a gradient quality where energy gradually becomes more negative until it reaches the point of "decay" Or Entropy.

The very centre most point of the flux tubes/ W- & W- Bosons/gluons are dipicted as Z Bosons. Their colour is black becuase they represent an Ever present micro neutral zone, a vacuum or micro black hole. This micro black hole has 0 value and 0 spin properties. Both positive head & negative Tail ends, boarder the Z Boson midpoint.

Quark Era/Epoch

We know that two gluons/ flux tubes create a Quark. The gluon flux tubes have 2 Neutrino heads and 2 Anti-Neutrino tails between them. Only 3 of these get used when a Quark forms. There is always either a Neutrino head or an Anti-Neutrino tail spare or 'lonley'. This "lonley" body part is depicted as blue. Part of It's 'body' belongs in the Quark but its counterpart head or tail does not and becomes part of the next forming Quark. This creates a chain or string like formation. The energy of forming Quarks is carried along the gluon / flux tube/ W- & W+ Boson structure and thus color flavors are seen. These coulor/flavors are found by the Nucleon- Deuteron model to be tiny oscillations caused by the energy flow though the positive Neutrino head and negative Anti-Neutrino tail states. Although they are not thought of as actual colours, they might contain within them a colour potential, which we shall look at later.

The Quark is now both particle and wave. Its wave form seems to be caused by Electro-Weakforce which runs along the dipole area, and Magnetic force which seems to run accross as a result of the Neutrinos attraction. When they pair up to form Quarks. The result should be an Electro-Magnetic wave / particle, which should give rise to the first oscillations in the universe.

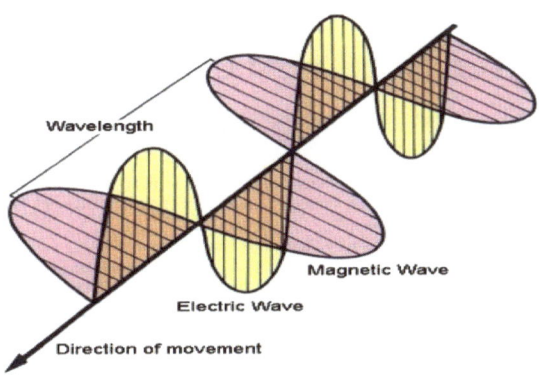

The Nucleon-Deuteron model reveals that Gluons/flux tubes/ W- & W+ Bosons are the medium through which information is carried or "transmitted" between Neutrinos & Quarks.

Science says that in an atom only 1% of the mass comes from Quarks. Science says that they might "steal" energy from virtual particles. The Nucleon-Deuteron model shows these "virtual" particles are Anti-particles. In this model they are Anti-Neutrinos which combine with the gluon flux tube/ W- & W+ Boson system and help to make up the mass of the atom. The Higgs field has also been found in the Nucleon-Deuteron model it seems to exist to give stability to the structure. This will be discussed during the "Higgs Sub-era" during which it should have formed.

The Z Boson & The Arrow Of Time.

The Nucleon-Deuteron Model finds that the Z Boson should lie at the midpoint of the Gluon flux tube/ W- & W+ Boson bodies. The Z Boson is a neutral vaccum environment. According to the Nucleon-Deuteron model it should act as a transition zone almost like an Event horizion in a Black hole stucture. Once (+1) energy passes through this zone the energy should decay to become negative (-1) energy. This (0) zone/ Z Boson seems to be the cause of Beta minus decay in atoms.

This idea is supported by the Passage of time theory, which was first proposed in 1988 by Seth Lloyd a graduate student at the cambridge university. At the time science was not ready for such a radicle theory and was for a while it was ignored. Now this theory is used in quantum computing. The theory describes how quantum entanglement causes the arrow of time. Lloyd described particles as being 1s and 0s. We now call this binary. The theory uses a cup of coffee to explain the passage of time. The particles in a cup of coffee would entangle with the air particles. This constent interaction would cause the particles to lose their origional identities which are deturmined by their spins of 1 or 0.

The whole particle system as a whole would Eventually contain all of the information leaving the individual particles without personal information. This means that all particles would reach a constant state of equilibrium. It is now thought that this must happen with all particles in a gradual energy dispersal. This Entropy is the disorder that the universe is going towards. It is thought to be the arrow of time. The theory also says that energy can become diffused but it is never compleatly disappears.

58
Quarks, Decay & The Arrow Of Time.

The Nucleon-Deuteron model shows that the arrow of time is governed by the rate of decay. The decay is facilitated by the gluon flux tube W- & W+ Boson stucture. The gluon flux tubes / W- & W+ Bosons have a gradint property where they pass through the Quark (up)/ Anti-Quark (down) pair. The vacuum action reduces when it reaches the center points. The gluon flux tubes/ W- & W+ Bosons are the bodies of the Neutrino pairs while the Neutrino heads and Anti-Neutrino tails become the Quark points or flavors.

Following the law of super-symmetry, the same must happen from the (-) tail end of the Neutrino to the centre. This is caused by the law of attraction. So we have a gravitational force and Nuclear Weak force acting on the particle pair. Each end has it's own angular momentum, creating a magnetic moment for the pair. This process leads to the decay of particles and the birth of new particles.

While the negative Anti-Neutrino Tail (-1) end is entangled with its positive Neutrino head via the gluon/ flux tube, the two polar ends are constenly trying to pull themselves closer together to reach the optimal state of oneness or equalibrium that they first experinced as a Majorana particle. This might be felt as a gravitational force, which causes the positive Neutrino head's (+1) enrgy that has matter potential to be pulled or attracted to the negative Anti-Neutrino (-1) tail end. Both the energy of the Neutrino head and Anti-Neutrino tail become attracted closer together until the (+1) Neutrino head's energy meets the concentration of the vacuum action which is the 0 mid-point Z Boson. This Z Boson is the Gluon flux tube/ W- & W+ Boson event horizon. Along its journey the (+1) head's energy is gradualy decaying, which would appear to be equilibriating /
entropy due to entanglement. Upon reaching the event horizon of the gluon fluxtube /vacuum/ micro black hole/ Z Boson, the (+1) head's energy would pass though the Z Boson /(0) vacuum and cross into the (-) zone. To the observer it would appear that the poitive Neutrino head energy (+1) had decayed from the physical dimension, or that it had reached a state of equilibrium and so can no longer be detected as having the properties that it once had.
We know that energy never disappears completely. The energy now exists in the (-) dimension in a united whole state of equilibrium where it no longer decays. The energy has become Anti-Energy or Dark Energy.
Due to the difficulties involved in detecting Dark Matter and Dark Energy we are led to believe that it is a strange substance that is somehow missing in atoms. However it is only undetectable to our frequencies and is not missing at all.

We know that a black hole draws a spinning stream of matter into its center. If we apply this to the Nucleon-Deuteron model then it could be that the Neutrino's body of energy gluon flux tube W-& W+ Bosons may also be spinning. The spinning might be responsible for the left and right handedness of the Neutrino heads and Anti-Neutrino tails. This could be because of forces being applied which draw the body- gluon flux tube/ W-& W+ Bosons inward and upward. During this process the neutral (0) center is trying to keep everything together by creating a force to pull it back into itself. As it does this the energy eventually & gradually gets drawn back to the center to reach the optimal equilibrium state of (0). This is what seems to happen to the first Primordial Father / Primordial Mother Neutrino pair inside the primitive Nucleon of Hydrogen The decay seems to spread through the Neutrinos in the Nucleon. This is what seems to cause the Nucleon inside a Hydrogen atom to decay into a Proton via Neutrino production. We measure time through change and decay. This seems to be the place where time as we know it might originate. Decay seems to happen to all matter in the universe in the same way. All matter should decay at a subatomic level, back through the (0) state through its Neutrino W- & W+ Boson flux tube and Z Boson /mini black hole system. Matter has been instructed and programmed to do this by the Majorana particle, who only wanted to pull itself together.

The Nucleon model suggests that this process could be the cause of ageing in all living beings. This (0) point seems to have stayed at the center of the original Majorana particle when it changed shape into the first Primordial Father (-1) & Primordial Mother (+1) Neutrino pair. If all atoms contain Neutrinos and all living beings contain atoms. We must all be bound by the inherited trait of the Majorana particle to decay or age.

Neutrino Pairs & Space-Time

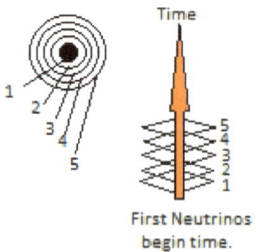

The circle with the rings around it, is a traditional diagram of space. The circles represent space where matter exists.
For the purpose of this model, on this diagram Point 1 represents the first Neutrino pairs.

Point 2 represents the next pairs and so on.
When we introduce the dimension of time we see that it seems to move the matter or particles forward. Each square represents Neutrino particles. The numbers represent the different states or stages in time. Beginning at state 1 where the first Neutrinos form, until the final release of the Electron and Positron at point 5. Change of state seems to cause the process which we interpret as time.

More about Electro-Weak Era/Epoch

Electro-Weak epoch began at 10^{-36} seconds and lasted until 10^{-12} seconds. The Temperatute at the beginning of this era was 10^{27} Degrees Kelvin. This cooled to 10^{15} Degress Kelvin by the end of the Epoch.

The nucleon-Deuteron Model finds that the length / wavelength of a Majorana singularity particle must be 2.0000×10^{-35}m
When measuring particles we do not usually measure what we cannot detect. Therfore the mass and other properties of Anti-particles are not usualy calculated. For example we would usually measure the properties of a Photon but the Anti Photon which is undetectable is not calculated. For consistency in measurements, the wavelength of the Majorana Neutrino pair will be calculated using the same principle, although we should remember that anti particles do exist and have properties which should mirror their particle.

During Quark epoch science says that this is where Symmetry breaking occurs. This is the point where particles no longer "seem" to be their own Anti-paricles. Each particle "apears" to have a seperate Anti-particle.

Traditional science says that an up Quark (uud) has a mass of 0.0002 Giga electron volts c2(Gev/c2)
Electric charge of 2/3
Spin 1/2 of unity.
A down Quark (ddu) has a Mass of 0.005 Gev/c2
Electric charge of -1/3

The Nucleon-Deuteron model puts the up and down (UDU)elements in a different order based on the order that Majorana Neutrinos seem to arrange themselves during Quark formation.

Traditional science says that the Higgs field interacts at this point. However the Nucleon-Deuteron model suggests that the Higgs field has no way of coming into being at this point because there is no need for it just yet. This seems to have been produced a little later on at around 10^{-2} sec.

The Nucleon-Deuteron model finds that the mass of the Quarks inside a primative free Neutron might be considerably lighter than it would be inside a Quark of a bound Neutron that is inside of regular nuclei. This is because at this primitve stage these Quarks which form the Neutron, have not developed the Higgs field. It is not until the primitive free Neutron decays into a Proton taking energy with it, through Neutrino production, that some of this conserved energy is able to be used for stability of the Proton. The conserved energy will be utilised for Higgs formation.

(Much later on during Beta plus decay in Deuterium formation, the Proton joins with another Proton. Then one Proton turns into a regular bound Neutron. As some nuclear binding energy is released, the changing Proton takes with it the Higgs field. Therefore this new generation of Neutron has a Higgs field and its Quarks seem to have more mass. This new generation of Neutron which formed in Deuterim is very different from the primitive primordial Neutron which began in Hydrogen.).

Equations for this epoch are as follows:

The up Quark contains within itself an up or (+) Majorana Neutrino followed by a down (–) anti Majorana Neutrino followed by an up (+) Majorana Neutrino.
So (+) plus (–) plus (+) = (+) because (+)&(-) cancel each other out so (+) is left over. Therfore the Quark is an up Quark and positivly charged (+). Two 3^{rds} of the Quark are positively charged because it contains two (+) Majorana Neutrino halves of dipole pairs. While one 3rd is negativly charged because it containes one Negativly charged anti Majorana Neutrino half of the dipole pair.

The Down Quark contains within itself a down or (-) Majorana Anti Neutrino followed by an up (+) Majorana Neutrino followed by a down (-) Majorana Anti Neutrino.

So (-) plus (+) plus (-) = (-) because (-) & (+) cancel each other out so (-) is left over. Therfore the Quark is a down Quark and negatively charged (-). Two 3^{rds} of the Quark are negatively charged because it contains two (-) Majorana Anti-Neutrino halfs of dipole pairs. While one 3rd is positivly charged because it containes one positivly charged Majorana Neutrino half of the dipole pair.

Electro-Magnetism causes the Neutrinos / Anti-Neutrinos to line up in opposite directions so that (+) attracts (-) and (-) attracts (+). In the model the New Neutrino (+) Head will be called Evelyn. The Anti-Neutrino (-) Tail will be called Adam.

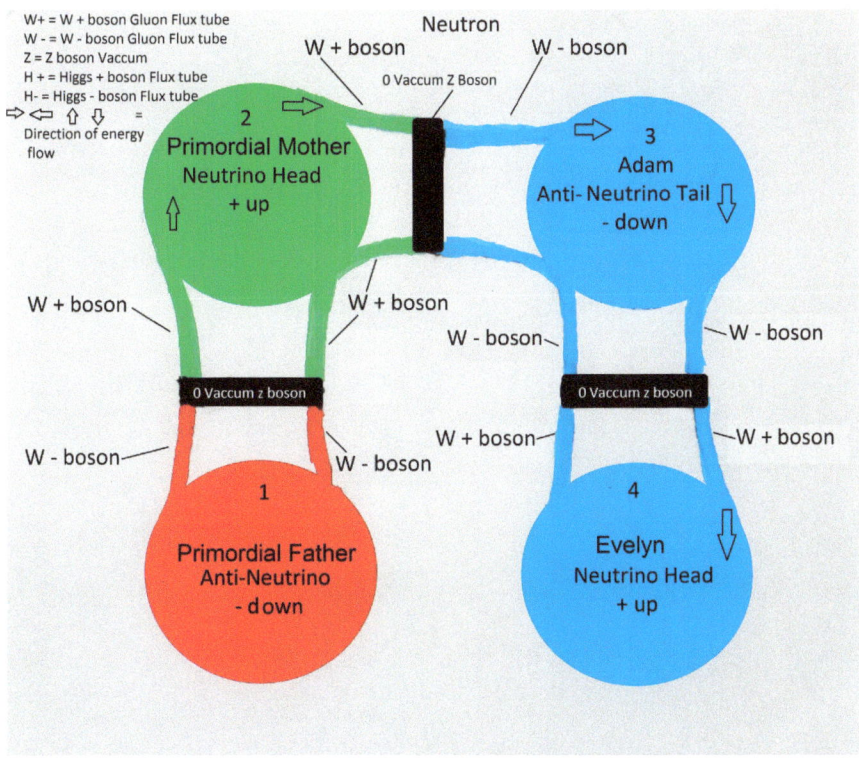

Adam & Evelyn create the first down Quark.

It is possible to conclude that this emission of another generation of Neutrino/ Anti-Neutrino pair was the first evidence of particle evolution and the 'replication system' of particles. It seems to mark the beginning of an "innate memory" leading Eventually to consciousness of life forms.

Here we can see that this entanglement should cause the formation of the first down Quark.

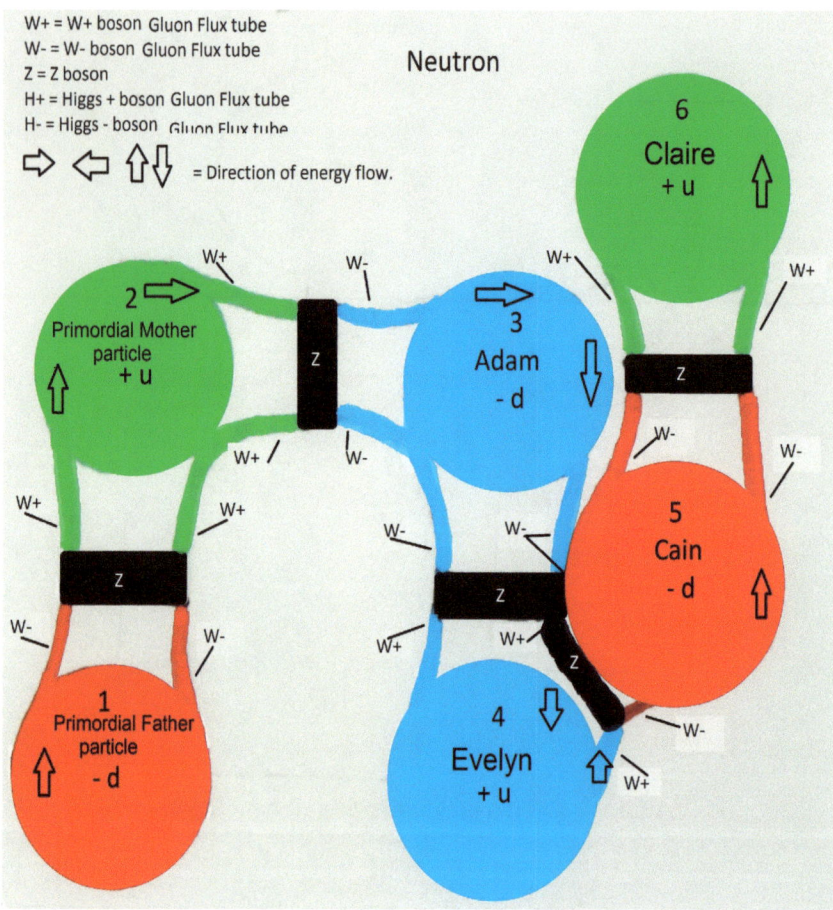

Cain & Claire are emitted, creating the first up Quark.

Evelyn's (+1) head, Cain's (-1) tail & Claire's (+1) head form the up Quark in the Nucleon. In this Quark the ratio of the (+1) Neutrino heads is higher than the ratio of the (-1) tails. Therfore the Quark is an up Quark. = UDU.
The Neutron still needs one more down Quark to complete it. Next Claire (+1) & Cain (-1) birth/ emit another Neutrino/Anti-Neutrino entangled pair. This pair are Able (-1) & Abby (+1).

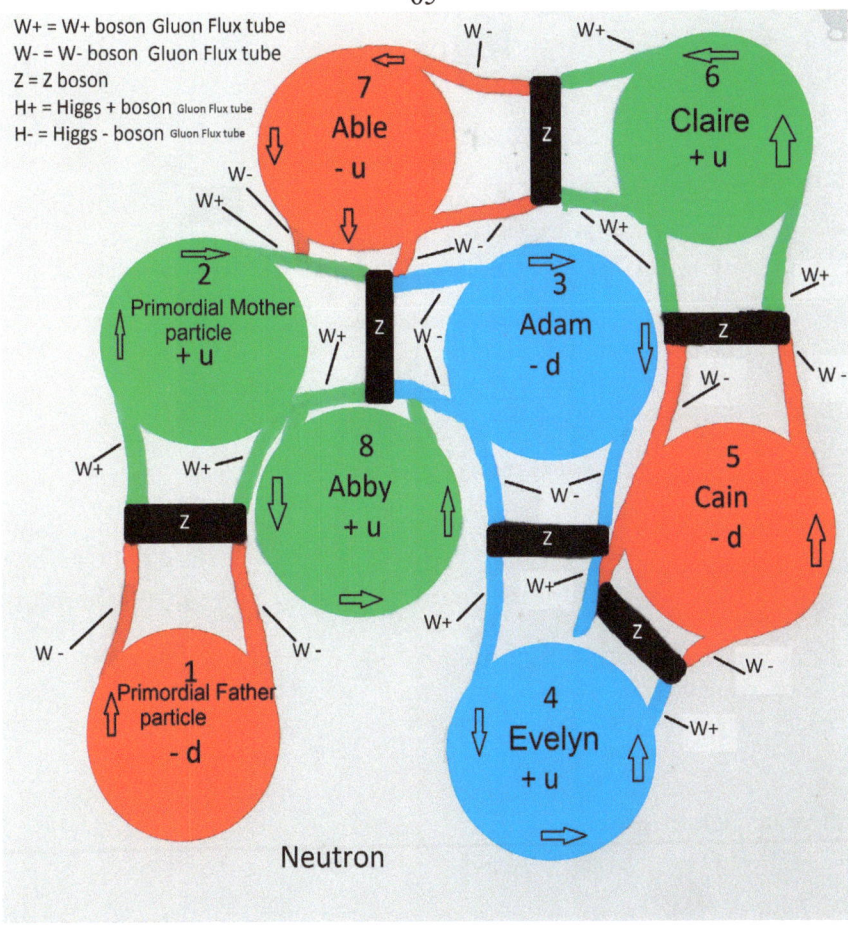

Able & Abby are emited, they begin the 2nd down Quark of the Neutron.

Able (-1) & Abby (+1) are not part of the 2nd up Quark family. They are orphaned or "lonely" and try to pull away causing stretching to their gluon / fluxtube body and also to the Neutron structure. The stretching may have been because Able & Abby are not tightly bound into the previous Quark via a head or tail. They are only bound because Able's tale is attracted to Claire's head. The new pair cannot form the last Quark alone.
Able (-1) & Abby (+1)emit another set of Neutrinos Brian
(-1) & Beth (+1).
The 1st Quark is dud = Down Quark.
The down (-) Tail end of the entangled Neutrino/ hold the potential to form Anti / Dark Matter and have right handed chairility.

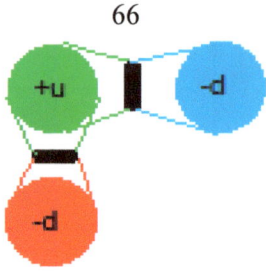

Down Quark.

The Up Quarks consist of 2 (+1) Neutrino up Heads and 1 (-1) down Anti-Neutrino Tail, which are entangled by W+ & W- Bosons / Gluon Flux tubes that form the "bodies" which join them. The overall flavor/ state of the Quark is Up because the ratio of up (+1) Neutrino heads is higher than the down (-1) Anti-Neutrino Tail ratio.
The (+) & (-)cancel each other out. Leaving a (+) therfore the Quark is a positive up Quark.
The Quark is udu = Up Quark.

The Up (-) Head end of the entangled Neutrino pair hold the potential to form ordinary Matter and left handed chairility.

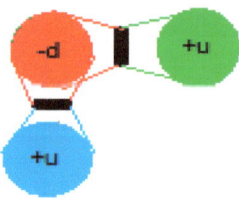

Up Quark

Science says that this is where symmetry breaking comes into play. Where Particles no longer seem to be their own Anti-particles but have separate Anti-particles. However there does not seem to be a <u>natural</u> Anti-Quark as a separate particle that forms alongside the up & down Quarks.

For example the Nucleon has contained within it both the positive and negative aspects of itself which comes from the Neutrino configuration within the Quarks. The Proton functions in the same way. Likewise there does not seem to be natural Anti-Neutrons or natural Anti-Protons as separate particles during Quark formation. The Anti-fractions of these particles are contained within the particle as a whole. The Neutrino heads and Anti-Neutrino tails form the Anti fractions of these particles.

Anti-Protons have been produced in the Hadron collider, however these particles do not seem to exist in the natural creation process which took place at the time of the "Big Bang". This is discussed furthur in the next section.

When this first Quark family is formed consisting of Primordial Father (-1) Anti-Neutrino Tail & Primordial Mother (+1) Neutrino head. It also contains Adam's Anti-Neutrino Tail (-1). This is the first Down Quark. However Evelyn's (+1) Neutrino head who does not fit inside the newly formed Quark family was left out and became the "lonely" (+1) Head. Evelyn's head was free from the Quark family so tries to pull away in hope of being 'adopted' by another Quark family which has yet to form. Evelyn's head remaines bound to her partner Adam through their shared Boson structure. The tail (-1) end has more mass because it exists on the negative side.

Summery Of The Equations For The Quark Era/Epoch

Equations for the Quark epoch :
C= speed of light 2.99792458×10^8 (metres per sec)
MSS = Majorana singularity speed/ velocity (metres per sec)
2.9979×10^8 m/sec
MNPS= Majorana neurtino pair speed 2.9979×10^8 m/sec
√Gh= plank time $<1.35 \times 10^{-43}$ (seconds)
pl = plank length 1.6000×10^{-35} (mm)
h= planks constant
T= temperature (Degrees Kelvin)
t= time (seconds)
λ= wavelength (mm)
f= frequencey (HZ)
E= energy (Joules or Gigga electron volts)
v= velocity (metres per sec)

To find the wavelength through-out the Quark epoch:
At 1.8125×10^{15} k the wavelength was 1.6000×10^{-18} mm
At 1.8125×10^{14} k the wavelength was 1.6000×10^{-17} mm
At 1.8125×10^{13} k the wavelength was 1.6000×10^{-16} mm
To find the frequencey at the end of the Quark epoch:

$$F = \frac{MSS\ 2.9979 \times 10^8}{\lambda\ 1.6000 \times 10^{-16} mm}$$

Frequencey = 1.8737×10^{24} HZ

Now we have the frequencey of the state of the universe at the end of Quark epoch. From this we can work out its velocity by multiplying its wavelength by its frequencey.

$$V = \lambda\ 1.6000 \times 10^{-16} mm \times f\ 1.8737 \times 10^{24} HZ$$

velocity = 2.9979×10^8 m/per/sec

Again this shows that the velocity remained constant.

To find the Energy at the end of the period, we can use the velocity or Majorana Neutrino pair speed (MNPS):

$$E = \frac{h \times MNPS\ 2.9979 \times 10^8}{\lambda\ 1.6000 \times 10^{-16} mm}$$

Energy = 1.2415×10^{-9} J

or

$$E = h \times f\ 1.8737 \times 10^{24} HZ$$

Energy = 1.2415×10^{-9} J

To find the mass during the Quark epoch, we should use Majorana

Neutrino pair speed.

$$\text{Mass} = \frac{\text{Energy } 1.2415 \times 10^{-9} J}{\text{MSS } 2.9979 \times 10^{8^2} \text{ m/sec}}$$

Mass = 1.3814×10^{-26} kg

Because $E=MC^2$

E= 1.3814×10^{-26} kg x $2.9979 \times 10^{8^2}$ m/sec

E= 1.2415×10^{-9} J

70
Anti-Quarks, Anti-Neutrons & Anti-Protons.

Anti-particles as we know have the same properties as their particle counterpart. The difference is that their spins are opposite. Majorana particles such as the Majorana singulaity and Majorana Neutrinos are their own Anti- particles. They form as anapole & dipole type particles and so have a positive and negative aspect to their 'bodies' which seems to be the cause of entanglement.
When we move on to larger particles such as Quarks, Neutrons and Protons it is traditionaly thought that they are not their own Anti-particle and symmetry breaks here. Each is said to have a seperate Anti-particle.
At the hadron collider experiments have been carried out that have produced Anti-particles. These Anti-particles do not exist naturally but are easily manufactured in the collider. One such experiment creates Anti-Protons. Anti Protons are produced by taking a normal Hydrogen atom and stripping its Electron. This process is Ionisation. After this they are left with a Proton. This is done on a mass scale so that enough Protons can be produced to make beams. The Proton beams are smashed together at high speeds, which produce high temperatures. The result of two Protons colliding is that we get three Protons and an Anti-Proton.

The Nucleon-Deuteron model shows how the Proton might have formed to create an atom, due to Neutrino replication. If this is correct then when two Protons collide in the collider they should create three Protons and an Anti-Proton. This is because the collision should have forced the Protons and their Quarks along with their Neutrinos into a jeperdised state. The Protons structure should have been blown apart causing scattering. However the Nuclear Weak and Strong forces should be strong enough to re-structure the Protons and pull the constituate parts back into place. As with any event we should see some energy released.

Usually energy releases are in the form of Neutrinos. During the collision the last pair of Neutrinos in each Proton, George and Georgina should feel jeperdised and release energy in the form of more Neutrinos. They would make copies of themselves very quickly and as they form they would naturally and rapidly form into Quarks, then Quarks form Protons. They would become Protons not Neutrons because they came from the collision of Proton energy. So each Proton has released a Neutrino pair that is responsible for creating another and so on until another Proton is made. However one new Proton is an Anti-Proton. This is becuse if you have two initial Protons both spinning in the same direction, whilst colliding, the energy might get spun off and the emitted energy could be sent in the opposite direction. In this case that energy was the new Neutrino pair which rapidly creates more pairs all spinning in this new direction, as its parent pairs did. The result should be a whole Proton spining in the opposite direction to the other three. An Anti-Proton is formed. This would be a whole Proton spinning in the opposite direction.

Why don't we see Neutrinos turn into Protons during the experiment? In most experiments we see Neutrinos emited. Usually Electrons, Positrons, Neutrinos or Anti-Neutrinos can be detected.

In these cases the particles can exist in this state breifly as they are emited. During the Positron collision the George and Georgina's generation of Neutrinos should have enough energy to create more to replace the ones who have been striped away. Electrons and Positrons. However as the energy is released during the collision, the energy might find that it cannot orbit the Proton that it "expects" to find. It finds itself in an environment where its Proton has been blown appart and finds itself in jepardy. It cannot exist and orbit as an Electron/ Positron pair so it releases energy.

This energy should become a new Neutrino pair. This all happens too fast to observe and so it appears that a new pair of Protons have spontaniously appeared. The production of Anti-Protons in the experients at the Hadron collider have only recently existed. We do not usually see them in nature. During the "Big Bang" such collisions might have been possible, where Protons might have collided. In such circumstances the Anti-Particles should have become part of the Anti-particle or Dark Matter/Dark Energy system. However Anti-Protons did not seem to have been part of the natural Proton production process during the "Big Bang".

Each Proton conatins within it the positive and negative / Anti-aspects of both Neutrinos and Quarks. This makes Quarks, Neutrons and Protons entangled with Anti-Particles and thus their own Anti-parts during formation. A seperate Anti-Particle is only produced in experiments or collisions.

Hadron Era/Epoch

Traditionaly this Epoch is described as being from 10^{-6} Seconds to 1 Second after the singularity point began to expand. The Temperature is thought to have been at 10^{13} Degrees Kelvin at the start of the Epoch and cooled to 10^{10} Degrees Kelvin by the end.

The Nucleon-Deuteron model finds that four quite amazing processes seem to have happened in this Epoch, therfore it should be split up into sub-eras. Each process might have caused each sub-era to last aprox 1.000×10^{2} seconds. These Sub-eras are as follows:
    Neutron Production,
    Neutron / Beta minus Decay,
    Proton Production
    Higgs Field production.
Science says that Hadrons are produced alongside Anti Hadrons. So the Neutron would have an Anti-Neutron and the Proton would have an Anti-Proton.
However the Nucleon-Deuteron model shows that the Hadrons might contain within themselves their own Anti-Hadrons. Meaning that they are a type of Majorana particle. For the Nucleon-Deuteron equation we take the value of only one Neutron or a Proton without including a seperate Anti-particle.

Neutron Production

This Sub-Era began at 10^{-6} Seconds and lasted for approx 1.000×10^{2} seconds. The Temperature at this time was 10^{13} degrees Kelvin.
Science says that 938 electron volts were needed to form a Neutron.
A Neutron is neutral and contains 2 down Quarks and 1 up Quark. $-1/3$ $-1/3$ & $+2/3$.

Traditional calculations of the Hadron epoch calculate the Hadron and Proton as well as their Anti-particles. Therfore they multiply the Neutron and Proton by 2. However for the purposes of this theory only one particle will be calculated because the Neutron and Protons contain within them their own Anti-particles.

Summery Of The Equations For The Hadron / Neutron Production Sub-Era :

Equations for the Hadron/ Neutron production sub-era :
C= speed of light 2.99792458x10^8 (metres per sec)
MSS = Majorana singularity speed/ velocity (metres per sec)
    2.9979x10^8m/sec
MNPS= Majorana Neurtino pair speed 2.9979x10^8m/sec
√Gh= plank time <1.35x10^-43(seconds)
pl = plank length 1.6000x10^-35(mm)
h= planks constant
T= temperature (Degrees Kelvin)
t= time (seconds)
λ= wavelength (mm)
f= frequencey (HZ)
E= energy (Joules or Gigga electron volts)
v= velocity (metres per sec)

To find the wavelength through-out the Hadron/ Neutron production sub-era:
At 1.8125x10^13k the wavelength was 1.6000x10^-16mm
At 1.8125x10^12k the wavelength was 1.6000x10^-15mm

To find the frequencey at the end of the Hadron/ Neutron production sub-era:

$$F = \frac{MSS\ 2.9979 \times 10^8}{\lambda\ 1.6000 \times 10^{-15} mm}$$
Frequencey = 1.8737x10^23HZ

Now we have the frequencey of the state of the universe at the end of the Hadron/ Neutron production sub-era:. From this we can work out its velocity by multiplying its wavelength by its frequencey.

V= λ 1.6000x10^-15mm x f1.8737x10^23HZ
   velocity = 2.9979x10^8 m/per/sec

Again this shows that the velocity remained constant.

To find the Energy at the end of the period, we can use the velocity or Majorana Neutrino pair speed (MNPS):
$$E = \frac{h \times MNPS\ 2.9979 \times 10^8}{\lambda\ 1.6000 \times 10^{-15} mm}$$

Energy = 1.2415×10^{-10} J
or
E= h x f 1.8737×10^{23} HZ
Energy = 1.2415×10^{-10} J

To find the mass we should use the Majorana Neutrino pair speed.

Majorana Mass = Majorana Energy $\dfrac{1.2415 \times 10^{-10} \text{J}}{\text{MSS } 2.9979 \times 10^{8^2} \text{ m/sec}}$

Majorana Mass= 1.3814×10^{-27} kg

Because $E=MC^2$

E= 1.3814×10^{-27} kg x $2.9979 \times 10^{8^2}$ m/sec
E = 1.2415×10^{-10} J

Energy per mole = E 1.2415×10^{-10} J x NA
E per mole = 7.4765×10^{13} J

Neutron Decay: Beta Minus Decay

This sub-era seems to happen simultaneously with the Proton Production process. This process might happen at aprox 10^{-4} Seconds. At a Temperature of around 10^{12} Degrees Kelvin. This Sub- era also seems responsible for the first Parity symmetry violation in the universe.

Weak interaction (vertices)
1) Charged current interaction mediated by an electric charge, carried by W+ & W- Bosons= Beta minus decay.

2) Neutral current interaction mediated by neutral Z Boson.

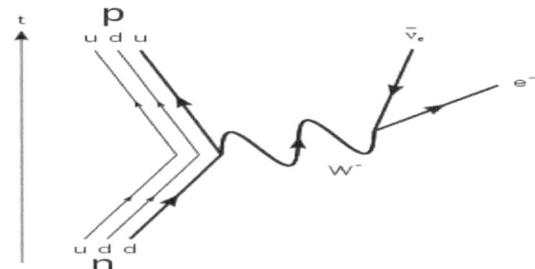

Beta decay is described as a down Quark in the Neutron emitting an Electron and an Electron Neurtino via a virtual W- Boson. The result is a Proton. The Nucleon-Deuteron model shows how this happens.
We begin with generation Brian Majorana Anti- Neutrino of the Neutron. This Anti-Neutrino is linked via flux tube W- Boson, Z Boson & W+ Boson to generation Beth Majorana Neutrino which will begin the Proton. The reason for this Proton production is that the Neutron Hadron is now closed off to Beth because it is full. Beth is now "lonley" and needs to connect to other similar particles to survive because Majorana Neutrinos cannot exsit outside of a Nucleon for very long before they decay. Beth & Brian upon being formed will almost immeditaly birth or emmit another like pair (Charlie + & Charles -). Adding to the Ever expanding universe. Beth's head does not fit into the Neutron she belongs in a new Hadron family which consists of Beth who is a positive Majorana Neutrino, Charles a negative Anti-Majorana Neutrino and Charlie a positive Majorana Neutrino. This New family begins the Proton. Obviously in the early universe it was likley that Neutrinos paired with the nearest pair, which might not have nessisarily been its own off spring.

When Beth finds herself outside of a closed Neutron a force might be felt between her & Brian, acting to seperate them. This force would cause them to stretch out away from each other slightly. This stretching might create a tempary strange Quark. However this strange Quark state would rapidly turn into a new state where Beth finds herself combining with Charles - & Charlie + forming the first up Quark of the Proton.
To begin with this primitive version of the Neutron is without stability. It does not seem to have a Higgs field and so decays when under the pressure of stretching out caused by Brian & Beths seperation, which was due to the Neutron being full. Therefore the Neutron begins to decay simultaiously with the production of the Proton.

To conserve energy, the energy of the Neutron is conserved through generation Brian & Beth via the W- & w+ Bosons. This energy of the Neutron should be utilised to become the Higgs field in the Proton. Now Beth & Brian are still linked via W-, W+ and Z Bosons. Brian appears to have decayed but because Beth has not decayed, his energy still exists, only in Anti or Dark energy form. His survival in this state is only possible because Brian was an Anti-particle to begin with and so would be unaffected by the decay. All the Majorana Neutrinos And their Anti-Neutrinos before him of the Nucleon would have had their energy travel through and become conserved in the Higgs field of the Proton. This Higgs field closes the gaps that otherwise exist in the Proton structure. The same gaps existed in the primitive Neutron which was the cause of its instability.

The Nucleon-Deuteron Model finds that the gaps appear during formation because when a Hadron forms its energy follows a certain path determin by its method of replication. The parts of the structure where energy does not travel is left unoccupied causing gaps which lead to instability. The Proton has no such problem because it now has extra energy which was conserved from the decayed Neutron. This conserved energy is sent to occupy these spaces and the stucture becomes incredibly stable.

This extra energy of the Higgs field adds a lot of Mass to the Proton because it holds conserved within it the mass/ energy of the original primitive primordial Neutron.
All other Neutrons after this point have a Higgs field because they are formed Via Beta plus decay, which is the decay of a Proton with Higgs field into a Neutron. The Higgs field is conserved as a Higgs field when the decay happens this way around.

Traditional Beta minus decay which involves the primitive Neutron decaying into a Proton says that the down Quark in a Neutron emits an Elecrton and an Electron Neutrino via a virtual W- Boson. The Nucleon-Deuteron model shows that when a Proton has been produced by replication and by using the conserved energy of the Neutron to form a Higgs field, it should emit an Electron and the Electrons Neutrino, via a W- Boson.

Summery Of The Equations For The Beta Minus Decay/Neutron Decay Sub-Era

Equations for the Beta minus Decay/Neutron decay sub-era :
C= speed of light 2.99792458×10^8 (metres per sec)
MSS = Majorana singularity speed/ velocity (metres per sec)
   2.9979×10^8 m/sec
MNPS= Majorana neurtino pair speed 2.9979×10^8 m/sec
$\sqrt{Gh}$= plank time $<1.35 \times 10^{-43}$ (seconds)
pl = plank length 1.6000×10^{-35} (mm)
h= planks constant
T= temperature (Degrees Kelvin)
t= time (seconds)
λ= wavelentgh (mm)
f= frequencey (HZ)
E= energy (Joules or Gigga electron volts)
v= velocity (metres per sec)

To find the wavelength through-out the Hadron/Neutron decay or beta minus decay sub era:
At 1.8125×10^{12}k the wavelength was 1.6000×10^{-15}mm
At 1.8125×10^{11}k the wavelength was 1.6000×10^{-14}mm
To find the frequencey at the end of the Hadron/Neutron production sub- era:

$$F = \frac{MSS \ 2.9979 \times 10^8}{\lambda \ 1.6000 \times 10^{-14} mm}$$
$$\text{Frequencey} = 1.8737 \times 10^{22} HZ$$

Now we have the frequencey of the state of the universe at the end of the beta minus/ Neutron decay sub- era:. From this we can work out its velocity by multiplying its wavelength by its frequencey.

V= λ 1.6000×10^{-14}mm x f1.8737×10^{22}HZ
   velocity = 2.9979×10^8 m/per/sec
Again this shows that the velocity remained constant.

To find the Energy at the end of the period, we can use the velocity or Majorana Neutrino pair speed (MNPS):

$$E = \frac{h \times MNPS \ 2.9979 \times 10^8}{\lambda \ 1.6000 \times 10^{-14} mm}$$
$$\text{Energy} = 1.2415 \times 10^{-11} J$$

or

$E = h \times f 1.8737 \times 10^{22} HZ$
Energy = $1.2415 \times 10^{-11} J$
To find the mass we should use the Majorana Neutrino pair speed.

$$\text{Mass} = \text{Majorana Energy} \frac{1.2415 \times 10^{-11} J}{MSS\ 2.9979 \times 10^{8}\ m/sec}$$
$$\text{Mass} = 1.3814 \times 10^{-28} kg$$

Because $E = MC^2$

$$E = 1.3814 \times 10^{-28} kg \times 2.9979 \times 10^{8}\ m/sec$$
$$E = 1.2415 \times 10^{-11} J$$

Energy per Mole = $E\ 1.2415 \times 10^{-11} J \times NA = 7.4765 \times 10^{12} J$

Proton Production
Sub-Era

As the Neutron Decays during Beta minus decay its energy is conserved in the production of the Proton.

The Sub-era might have began at 10^-3 Seconds and lasted until 1 Second after the expansion of the singularity point. This process might have happened as the Neutron was Decaying. The Proton seems to have got its energy from the Beta – Decay of the Neutron. Almost like an energy transfer or primitive evolutionary process.

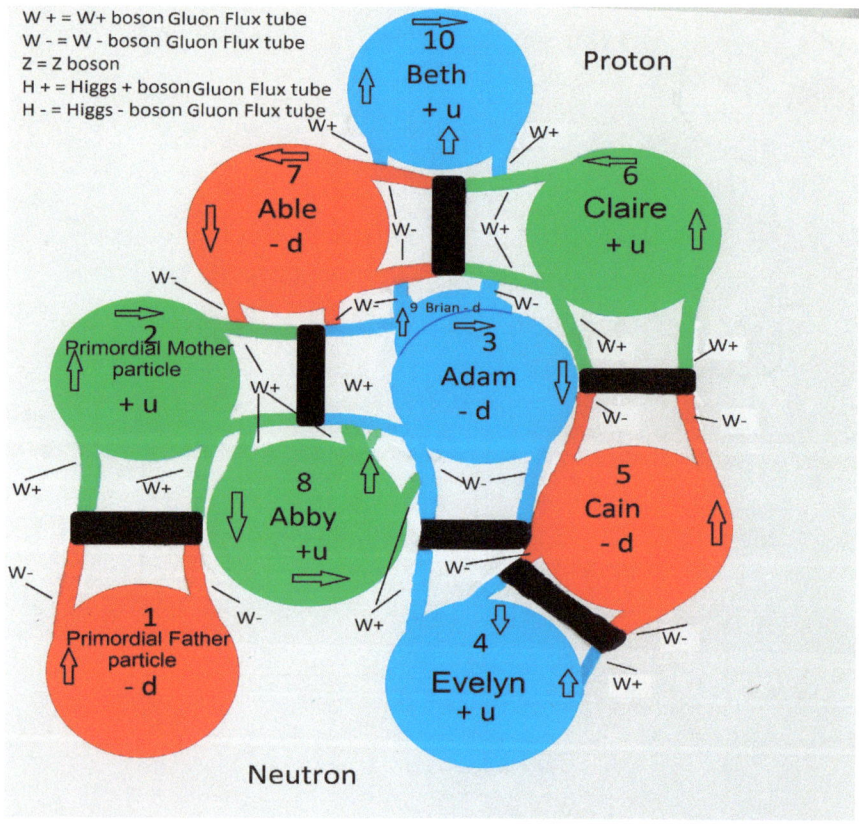

Brian & Beth are emitted. Beth begins the first up Quark in the Proton.

To re-cap : Able (-1), Abby (+1) & Brian (-1) form the last Quark of the Hydrogen Neutron. This Quark has a higher ratio of down (-1) Anti-Neutrino tails than (+1) Neutrino heads, therefore it is a down Quark.

Beth's (+1) head is left out of this last Quark family and the entire Neutron is closed to her. Beth's head carries the energy over into the production of the next part of the structure.

As Beth stretches out to do this the Neutron behind her rapidly decays. As mentioned earlier, Beth's stretching might have caused her to create a natural strange Quark. (Strange Quarks are usualy only seeen in lab experiments where a Quark is being pulled out of its Nucleon. The strange Quark has a short life and quickly decays or turns back into a down Quark. A natural strange Quark might be possible due to the stretched out state of the Hadron while a quantum leap such as energy tranformation from Neutron to Proton is taking place).

This stretching would have caused Beth and Brian to emit another Neutrino/Anti-Neutrino pair. This emmission of the Neutrino/Anti-Neutrino pair might also be the cause of the instability which caused the decay of the Neutron.

Almost instantly the first Primordial Father & Primordial Mother Neutrino / Anti-Neutrino pair decay. The energy of the Primordial Mother head may have been pulled back towards its centre rapidly because it was in its "nature" to do this. It retains the ability which is remembered or inherited from it's Majorana form.

The Beta minus decay effects all of the Majorana Neutrino pairs in the Neutron in turn.
Although the decayed Neutrino heads & tails of the primitive Hydrogeon Neutron have passed on their energy, they are technically no longer occupying the "space" that they used to. However they are still depicted in the diagrams. This is to keep a sence of how the atomic structure progresses and of the geometry it forms. The decayed elements are now depicted with black rings around them to indicate the Beta minus decay they have undergone.

Traditionaly models show Neutrons and Protons occupying the same space co-ordinates as they decay into one another. Atoms are too small to actualy see this process. However the Nucleon-Deuteron model sugests that as the primitive Hydrogen Neutron decays, it's energy travels slightly to form the Proton. This is because it is being formed due to the Majorana reproduction process, which causes a projection of the off spring energy.

Beta Plus decay from Proton back to Neutron has different rules because it consists of a combination of decay & Neutrino flipping. The result is that Protons should turn into Neutrons on the spot as traditional models predict.

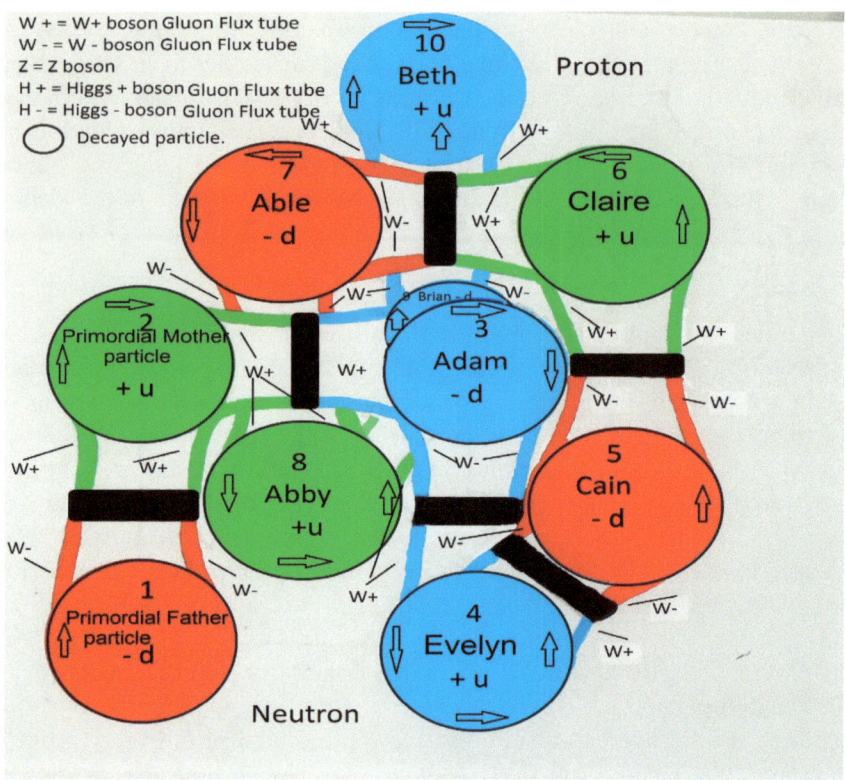

The Neutron begins to experience Beta minus decay.
The decay of the Neutron is seen in Hydrogen Atoms. The Primordial Father /Primordial Mother Neutrino /Anti-Neutrino pair are primitive and have not yet developed the ability to become stabilised by forming extra Gluon / Flux tubes to other Neutrinos in the structure or Higgs mechanism.

Beth's (+1) Neutrino head survives the decay although it is still attached to Brian's Anti-Neutrino Tail via a Gluon / Flux tube. Brian appears missing along with the rest of the Neutron. However Brian's tail end was already a (-1) Anti-particle with 1/2 spin of unity. It was existing in the negative zone before the decay of the Neutron. Therefore his decay should not have effected his entangled counterpart, Beth's (+1) Neutrino head.

The W- & + W Gluon /Flux tube entanglement between a Neutrino head, body and Anti-Neutrino tail cannot be broken Even though the Neutron stucture appears to decay or quantum leaps into the negative zone.
Before the decay of the Neutron is complete Brian (-1) & Beth (+1) emit another Neutrino set Charles (-1) & Charlie (+1).

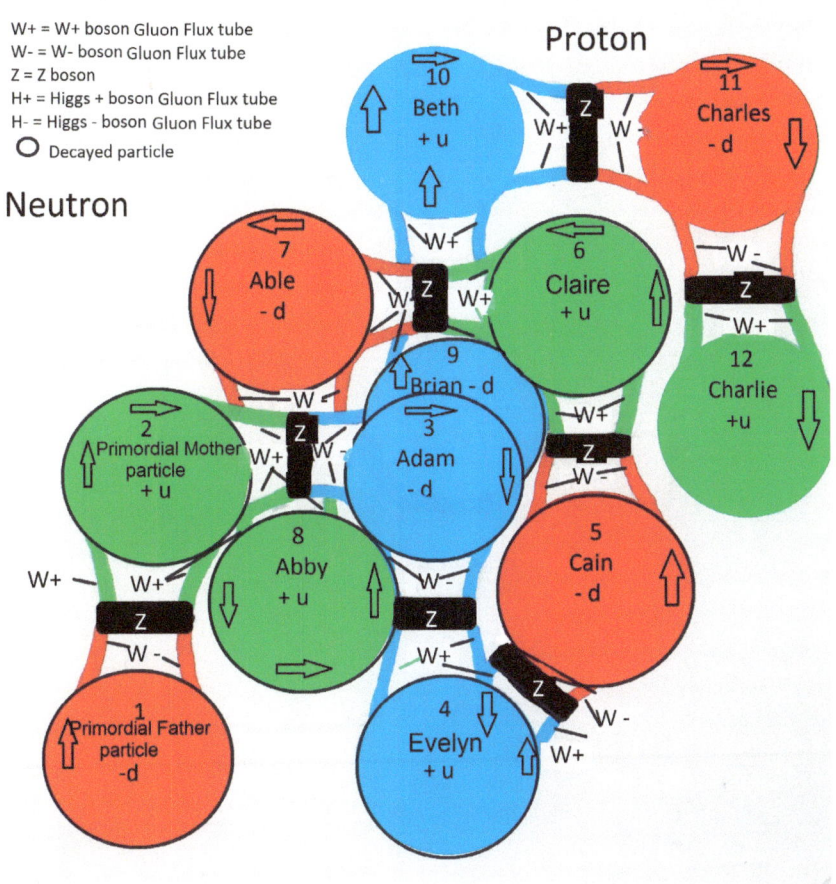

Charles & Charlie are emitted. The first up Quark of the Proton is compleate.

Beth begins the formation of the Proton with the new set of Quarks as the Neutron is simultaneously decaying. The Energy is passed through from Beth's head (+1) into Charles tail (-1) then into Charlie's (+1) head. This becomes the first Up Quark in the Proton of the Hydrogen atom. The Quark is an up Quark because it has a higher ratio of (+1) up heads than (-1) down tails. = udu.

The "lifeforce" Energy is now concentrated in the Proton. With the Neutron's Neutrinos missing, there is a hole in the energy "matrix" of the cube which makes the structure primitive and unstable. The reason that the Hydrogen atom cannot progress and evolve at this point is due to this instability. The energy continues to pass through the Proton while it is being formed.

Charles (-1) tail and Charlie's (+1) head birth another Neutrino pair, Dan (-1) & Diana (+1). At the moment of Neutron decay, the two pairs resemble a Meson, which is a 2 Quark structure.

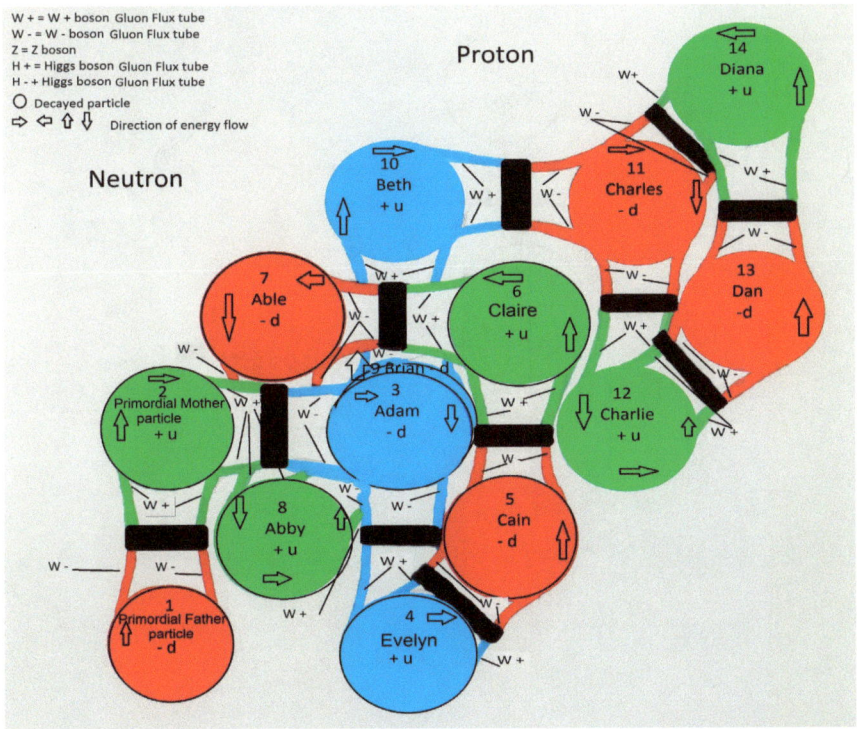

Dan & Diana are emitted. They begin the down Quark of the Proton.

Dan and Diana emmit another Neutrino /Anti-Neutrino pair, Eddie (-1) & Edna (+1) who carry on the energy.

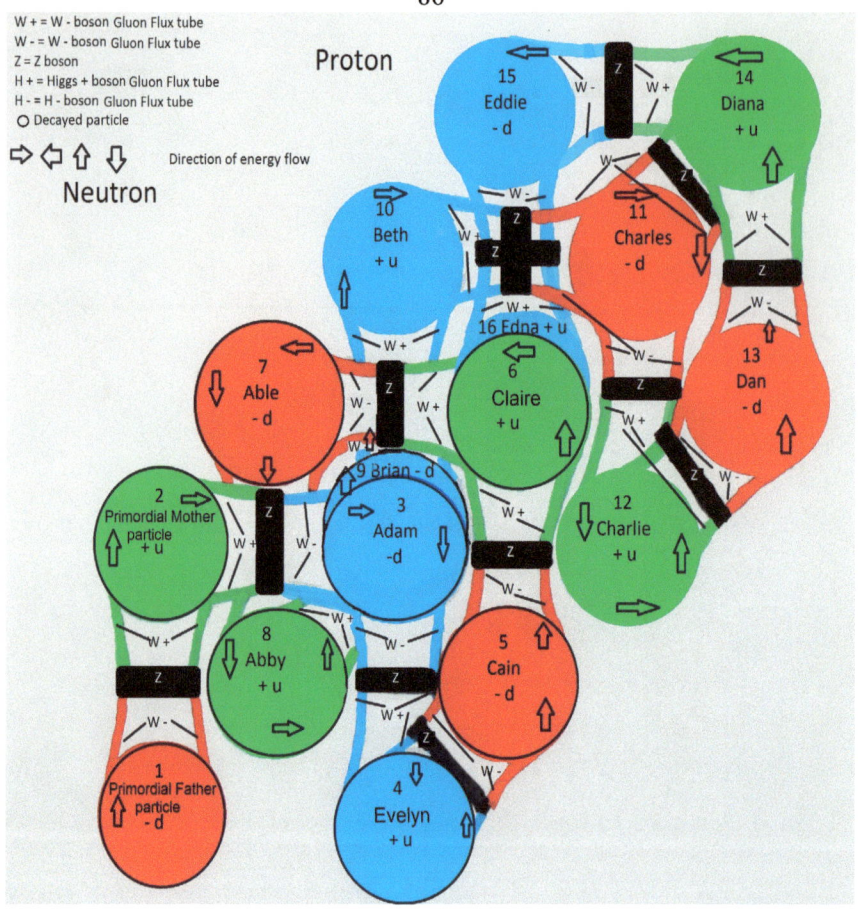

Eddie & Edna are emitted. Eddie completes the down Quark while Edna begins the last up Quark of the Proton.

Generation Dan (-1), Diana (+) & Eddie (-1) form the second Quark in the Proton of the Hydrogen atom. This Quark is down because its Anti- Neutrino tail ratio (-1) down is higher than the Neutrino head (+1) up. The Quark = dud. Edna is left out of this Quark and becomes the "lonely" head (+1). Eddie (-1) & Edna (+1) birth another Neutrino set Freddy (-1) & Frayer (+1).

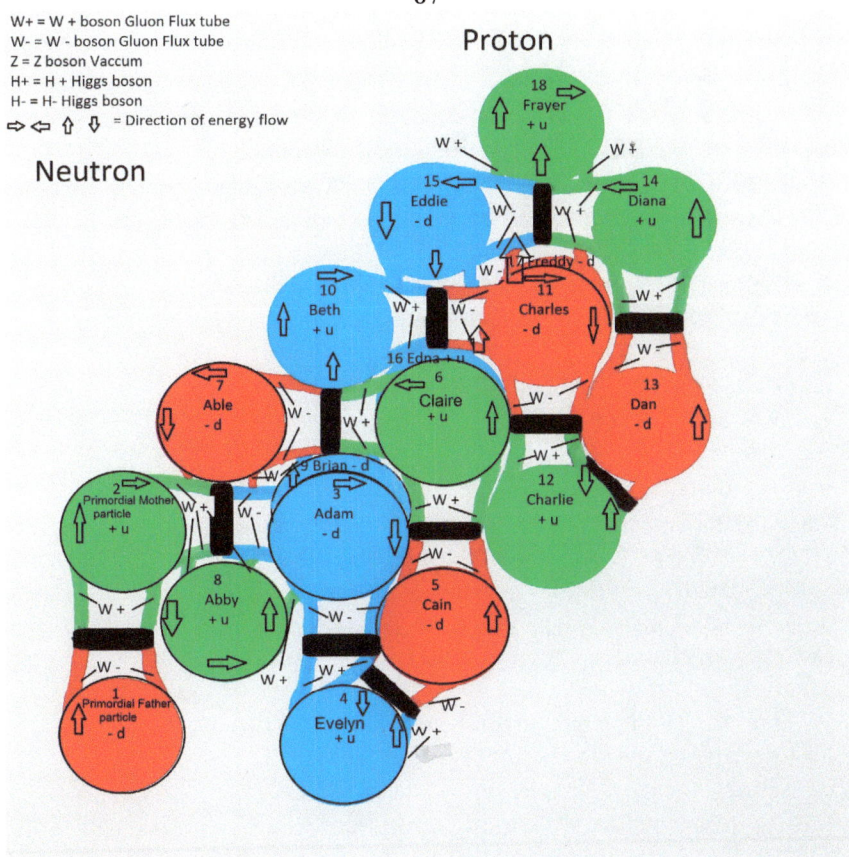

Freddy & Frayer are emitted. They compleate the final up Quark of the Proton.

Edna's Neutrino head (+1) entangles with Freddy's Anti-Neutrino (-1) tail & Frayer's (+1) head. The trio form the last Quark in the Proton of the Hydrogen atom. This Quark's (+1) Neutrino head ratio is higher than the (-1) Anti- Neutrino tail ratio. Therefore the Quark is an up Quark. = udu. Freddy and Frayer emit George and Georgina.

Summery Of The Equations For The Hadron: Proton Production Sub-era

Equations for the Hadron/ Proton production sub-era :
C= speed of light 2.99792458×10^8 (metres per sec)
MSS = Majorana singularity speed/ velocity (metres per sec)
   2.9979×10^8 m/sec
MNPS= Majorana neurtino pair speed 2.9979×10^8 m/sec
$\sqrt{Gh}$= plank time $<1.35 \times 10^{-43}$ (seconds)
pl = plank length 1.6000×10^{-35} (mm)
h= planks constant
T= temperature (Degrees Kelvin)
t= time (seconds)
λ= wavelength (mm)
f= frequencey (HZ)
E= energy (Joules or Gigga electron volts)
v= velocity (metres per sec)

To find the wavelength through-out the Hadron/ Proton production sub-era.

At 1.8125×10^{11}k the wavelength was 1.6000×10^{-14}mm
At 1.8125×10^{10}k the wavelength was 1.6000×10^{-13}mm

To find the frequencey at the end of the Hadron/Proton production sub- era:

$$F = \frac{MSS\ 2.9979 \times 10^8}{\lambda\ 1.6000 \times 10^{-13} mm}$$

Frequencey = 1.8737×10^{21} HZ

Now we have the frequencey of the state of the universe at the end of the beta minus/ Neutron decay sub- era:. From this we can work out its velocity by multiplying its wavelength by its frequencey.

V= λ 1.6000×10^{-13}mm x f1.8737×10^{21}HZ
  Velocity = 2.9979×10^8 m/per/sec
Again this shows that the velocity remained constant.

To find the Energy at the end of the period, we can use the velocity or Majorana Neutrino pair speed (MNPS):

$$E = \frac{h \times MNPS\ 2.9979 \times 10^8}{\lambda\ 1.6000 \times 10^{-13} mm}$$

Energy = 1.2415×10^{-12} J
   or

$E = h \times f\, 1.8737 \times 10^{21} HZ$

Energy = $1.2415 \times 10^{-12} J$

To find the mass we should use the Majorana Neutrino pair speed.

$$\text{Mass} = \frac{\text{Energy } 1.2415 \times 10^{-12} J}{\text{MSS } 2.9979 \times 10^{8^2} \text{ m/sec}}$$

Mass = $1.3808 \times 10^{-29} kg$

Because $E = MC^2$

$E = 1.3808 \times 10^{-29} kg \times 2.9979 \times 10^{8^2}$ m/sec

$= E\, 1.2415 \times 10^{-12} J$

Energy per mole = $E\, 1.2415 \times 10^{-12} J \times NA = 7.4765 \times 10^{11} J$

The Nucleon-Deuteron model suggests that the Proton forms in the following way:

George and Georgina Eventually emit the Electron and Positron.
Beta – decay says that the Electon and its Positron are emitted via a W-Boson. This says that the energy travels from Georgina and released as the Electon/ Positron pair through generation Georges W- Boson.

The Electron is therfore a left-handed particle and negative. The Electron Neutrino is said to also have a left handed chairality because it is formed from the left-handed Electron. Amino acids and molecules have been obsreved to have left-handed chairality. The left-handed chairality seems to make them physical. If they were to have a right-handed chairality they would not form as amino acids & molecules as we know them.

Cosmology says that by the end of the Hadron epoch the Ratio of Neutrons to Protons was close to 1:1 as the temperature dropped there was an equilibrium shift as Protons lower their mass to aprox 7 mev 1 second. This resulted in a freeze out ratio of 1/6. of free Neutrons. These free Neutrons were unstable with a meanlife of 880 sec. By the end of the epoch they had decayed so the ratio was 1/7 where almost all the Neutrons had fused without decaying and combind to form Helium 4 and some Deuterium also remained. However It might be that Neutrons were free until they Decayed into Protons. The Protons combined to form Deuterium which must have taken many years. Deuterium forms as a result of Beta plus decay. Then most of the Deuterium combins to form Helim 4.

It seems that the Neutron has an innate programming which was inherited from its parts. The programing began with the Majorana, it was conserved in the Neutrino/ Anti Neutrino pairs , then futrher conserved in the Quark structure. The programming is still present in the Neutron and Proton. It does not stop there it is also conserved in the Electron/ Positron and in the Photon/ Anti Photon entangled pairs. This is why the Proton can change its state back into a Neutron upon beta plus decay and why an Electron and its Positron can drop down in state. This is also why the Photon can also change its state. They all contain within themselves the memory programming of their previous states to be called upon at a later time.
This is also seen in nature as a part of the evolutionary process where a creature can evolve a body part away then over time evlove this body part back again. This effect has been seen in whales. Where a whale autopsy showed that although all life originated in the sea as did the whale it then evolved and grew hind legs and walked upon land for a time many years ago. The whale then decided to go back into the sea and its hind legs have

become folded into its body stucture and it has grown a tail around them. The whale still has the genetic coding to have hind legs although its evolutionary process has caused it to favour the tail instead due to its envirionment.

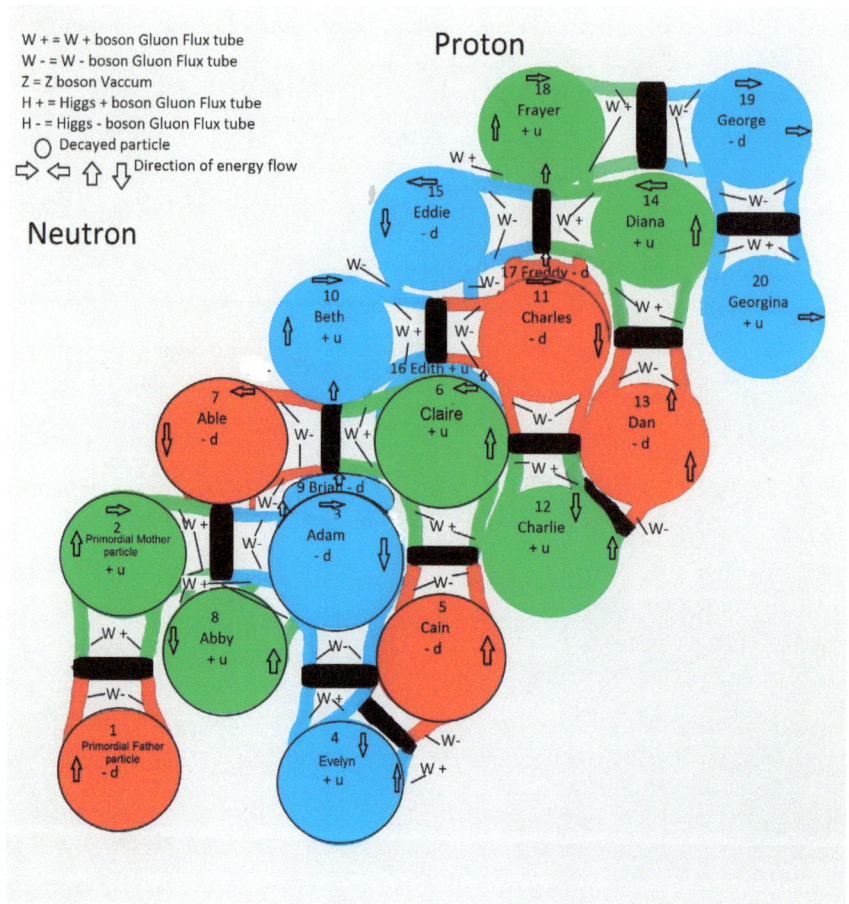

George & Georgina are emitted.

George & Georgina do not form a Quark but become the Electron Neutrino & Positron. Their energy breaks free to orbit the Proton. As they do this they gain energy and mass.
The creation energy runs along this path via the gluon /flux tubes, W+ & W- Bosons.

Higgs Field Production Sub-Era

This process also seems to have been happening simultaiously with the Proton production sub-era. The Higgs Field/mechanism seems to have formed as part of the Proton in the form of Bosons. It should have finished forming at the end of the Proton era at 1 second after the expansion of the singularity point. The Higgs field should have formed at temperatures that were aprox 10^{11} Degrees Kelvin to 10^{10} Degrees Kelvin.

The Nucleon structure should have formed a gluon type flux tube system as a stability measure from binding energy. These extra flux tubes might also create a force which exists to keep the structure together. It may be responsible for strong force and also provides mass to the Nucleon. These extra flux tubes might be the Higgs field. The Higgs field is thought to produce the Higgs Boson due to field excitement.

The Higgs Boson is said to have parity, 0 color charge so it might be directly involved with the Quark oscillations or colour change. It has 0 electric charge.
If the Boson type pathways of the Nucleon-Deuteron modle are the Higgs field, there would have to be two types. The Higgs field on the (+) side would have to be a + Higgs field while the other would be a - Higgs field. These are depicted as yellow.

On the 4th of July 2012 experiments by ATLAS & CMS at the Large Hadron Collider in CERN both confirmed the existence of the Higgs. They did this independently of each other. They report the Higgs as having a mass of aprox 125GEV/c2 which is 133 times the mass of a Proton. The Higgs is in the order of 10^{-25}kg.

The Nucleon-Deuteron model suggests that the Higgs Boson exists at a mass of 10^{-25}kg which is 125GEV. It suggests that the Higgs field which facilitates the Higgs Boson must be slightly smaller in the order of 10^{-30}kg. The Higgs fields energy can be seen in the binding energy of Deuterium. When the Higgs Boson appears in the Higgs field the energy rises due to exitation. The energy should rise to 10^{-25}kg to produce the Higgs Boson.

To explane how the Higgs field might have a mass in the order of 10^{-30}kg we must first look at the equation which explaines the binding energy of Hydrogen.

H= 1.00794u
Proton = 1.007276u.
We need to subtract the Proton mass from the total mass of the Hydrogen atom:
(H)1.00794 −(P)1.007276u = 0.000664u
This means that Hydrogen's mass equils more than the mass of the sum of its parts by 0.000664u which is equil to $1.1025986928 \times 10^{-30}$kg. According to the Nucleon-Deuteron model this extra nuclear binding energy is used to form the Higgs field which facilitates to Higgs Boson.

The Nucleon-Deuteron model shows that the Higgs field must have both - & + qualities. The + quality might be relatively easy to observe however the − quality might not due to it having an anti feild quality. Therfore we might only think to measure its + quality which shows up in the binding energy. Both are important because they both interact with each other via a z Boson which seperates the two states. To measure both + & - we must times the binding energy by 2.

The equation should be:
$(2)1.1025986928 \times 10^{-30}$kg
$= 2.2052 \times 10^{-30}$kg

Here we can see that the mass of the Plus & Negative Higgs field should be 2.2052×10^{-30}kg.
The value of the Higgs field is close to the value of the Higgs Boson itself at $2.240163951 \times 10^{-25}$kg.

The reason that only half the Higgs field value of $1.1025986928 \times 10^{-30}$kg is seen in the Hydrogen atom might be that the primitive Hydrogen atom does not need a full Higgs field because it only has a Proton. Therefore it appears that the Hydrogen atom has more binding energy left over than larger atoms. When we calculate the parts of a Hydrogen atom we count the Quarks. However we have not yet counted the energy that has been conserved within the Proton from the Neutron decay. The Neutrons energy should be present in the Proton in the form of the Higgs field which the Proton uses for stability. This then adds mass to the Proton. This could be why the Hydrogen atom containing only one Proton seems heavier than the sum of its parts. It is the only atom with this unique feature which is due to its primitve nature.

Larger atoms have both Neutrons and Protons. If we take Deuterium as an example we see that it has a Neutron and Proton. We know that Deuterium is formed from two Protons combining. Here Beta plus decay takes place. Where two Protons cannot exist adjoined in this manner because Protons have plus charge and would naturally repel each other. So one Proton changes or decays back into a Neutron by flipping it's Neutrino / Anti-Neutrino entangled pairs, due to repulsion. However these Protons have a Higgs field attached to them which they inherited from the left over binding energy of the earlier beta minus decay which helped to form them.

As we know energy is always conserved so the Proton must take its energy with it when it decays. The result is that it decays into a 2nd generation Neutron which should use the mass defect energy to form its own Higgs field. This Higgs field is important to the Neutron because it gives it stability and keeps it from decaying back into a Proton. The Neutron and Proton now both have the stability they need to survive and the strentgh to be built upon. The strength that the Higgs field provides is the reason that heavier more complex atoms can be built.

Therefore in the Hydrogen atom only half the Higgs field can be seen because a full Higgs field with both positive & negative aspects should be 2.2052×10^{-30}kg. This is divided between the Neutron and Proton of the Deuterium atom. Both Neutron and Proton contain positive & negative aspects of this field.

In larger atoms the full Higgs field should also become larger because they contain more Neutrons and Protons. The binding energies for different atoms will be different and this should directly effect the Higgs field.

The Higgs does not seem to form part of the path that energy flows along, it seems to be there purely for stability and due to energy conservation. It looks very similar to W- & W+ Bosons. It is connected to their network. When energy travels along W- & W+ Bosons which join the Higgs feild, the energy exites the Higgs field this is when a Higgs Boson is likely to be produced. The field energy is raised from (Higgs field + & -) 2.2052×10^{-30}kg to $2.240163951 \times 10^{-25}$kg. Although it is worth remembering that when we look for the Higgs Boson, we seem to be detecting its positive form and its negative aspect is not calculated. If we were only to look at the positive aspect of the Higgs field that we can see being exited it would be $1.1025986928 \times 10^{-30}$kg. The Higgs Boson rapidly decays.

The Higgs goes back to resembling a Higgs field which looks remarkebly like W & Z Bosons. This might be why experiments find that the Higgs Boson seems to decay into W & Z Bosons.

Beta Plus Decay In Deuterium Formation

As Protons become close they feel a repulsion. Strongforce, which is mediated by W+ & W- Bosons and the Higgs field, keeps them together. This repulsion causes the Quarks to react to each other. :
Protons =2 up & 1 down Quark.

The two up Quarks at the ends are in contact with 2 up Majorana Neutrinos. They cannot exist in this state of 1/2 spin next to 1/2 spin. So one Majorana Neutrino is forced to change its direction/state and becomes -1/2 spin, in the oppsite direction. This makes it a Majorana Neutrino/ this causes neighbouring down -1/2 Anti-Majorana Neutrino to change to up 1/2 Majorana Neutrino.

The next Majorana Neutrino detects this as it is being repelled. It changes to a down -1/2 Anti Majorana Neutrino and so on until all of the Proton's Majorana Neutrinos have changed state / direction of spin. The Majorana Neutrinos direction of spin of state effects the Quark flavors. So an up Quark will become a down Quark and a down Quark will become an up Quark. Therfore the Proton becomes a Neutron.

Higgs Boson & Field

The Nucleon-Deuteron Model finds that, during this process when the two Protons combine they find they no longer need all the outer connections of the Higgs field for stability. So this excess field is discharged or emitted as mass defect energy, Electons and Neutrinos casing the atom to become lighter than the sum of its Hadrons. The energy connections of the Higgs are used as binding energy.
.002388u which is aprox $3.9653699976 \times 10^{-30}$. This energy is the Higgs field.

The Higgs in Deuterium.

There is .002388u of binding energy lost in Deuterium formation. Therfore Deuterium appears to have .002388u less mass than the sum of its parts. The reason for this might be that in this case much of the Higgs field has decayed from its Proton form into the Neutron form.
So the Higgs minus .002388 = $3.9653699976 \times 10^{-30}$ kg.

The Higgs positive & negative field mechanism should = 2.2052×10^{-30} kg, which when exited by passing energy should form a Higgs Boson of $2.240163951 \times 10^{-30}$ kg.
As we can see the Hydrogen atom is the exeption to the rule of regular atoms. The reason for this descrepency seem to lie in the structure & nature of the Higgs field. We can see that the "lifeforce" energy of the Nucleon structure began with a single Majorana particle which "evolved" into a Neutrino for the purpose of "self preservation". Upon "feeling" jeopardized

because of stretching it developed the ability to replicate itself. This action gave rise to the "replication' system of the Nucleon. It is clear to see that the energy gets passed on from one generation of Majorana Neutrino to the next as the Nucleon develops. The Majorana Neutrinos energy seems to pass through up (positive) heads & down (negative) tails in an alternating pattern. This process causes a series of oscillations between the two states. To an observer this process can only be observed in the Quarks that these Majorana Neutrinos make up. Therefore it appears that the Quark itself is changing state / flavor or color/ frequency. The flavors/ colors/ frquencies or states of the Quarks would seem to an observer as being a net color. The colors used in this model are primary colors. The Quarks colors start with red then turn green through the spectrum to blue which indicates the "lonely" Neutrino head or Anti- Neutrino tail. Eventualy the energy passes through an atom to be emitted as a Photon & Anti-Photon of light. The light carries the primary colours/frequencies and when refracted the light reveales all of its hughes or frequencies as a rainbow.

The rainbow frequency potential was always contained within the structure which began with the first Primordial Father, Primordial Mother, Evelyn down Quark that formed.
The energy travels through the Nucleon maze structure creating a checkerboard effect.

The energy seems to travel through the Nucleon structure in a similar way to how an aircraft being the source of energy, leaves behind a smoke trail of information which indicates where it had been. If this aircraft had been concord it would have exceeded the speed of sound thus causing a sonic boom. Likewise the tiny Neutrino's energy is travelling so fast and decays with such an impact that it might cause a similar nuclear boom when energy falls below (0) point and returns to enropy though the gloun flux tube W- & W+ Bosons and the Z Boson, micro Black hole structure. A similar nuclear boom and mini gamma-like ray might be caused when the Gluon/ Flux tube field stretches and births/ emmits a new Majorana Neutrino entangled pair.

Up until now scientists have thought of Neutrinos as having 3 states. Professor Kevin McFarland is a spokesperson for the MINERvA Neutrino experiment, under the department of Energy's Fermilab at the Rochester University.
On 06/07/2016 their work was featured in an article in Symmetry magazine. The article was entitled "Travelling Neutrinos are in several states at once". Meaning that at any one time the "Neutrino is some fraction of all 3 flavors". They can be in their eigenstates of definate mass or in overlap states which have a "mixing" angle.

According to the Nucleon-Deuteron model, when this happens to the Majorana Neutrinos inside of Quarks the effect seems to be caused by the energy of the forming Nucleon. The Oscillating energy is actualy traveling through the positive Neutrino heads & negative Anti-Neutrino tails in an alternating pattern. At any one time there are 3 Neutrino heads & Anti-Neutrino tails in the Quark, displaying flavors / states or characteristics. The end result is the release of the last flavor or state of Majorana Neutrino to be released from the Proton. This Majorana Neutrino is outside of the Proton but is still entangled with it. To survive it must change orbit the nucleus in a shell. As it gains energy it moves up another shell away from the nucleus. Its state changes. This new form of Majorana Neutrino becomes the Electron and its Positron. They might be larger than the Majorana Neutrnos inside the nucleus because they have gained more energy.

Summery of Hadron Formation

The Nucleon-Deuteron model shows how 3 Quarks were formed in the Neutron of the Hydrogen atom. They were:

  dud=
 Down Quark (Primordial Father ,Primordial Mother ,Adam)
 udu = Up Quark   (Evelyn,Cain,Claire)
 dud = Down Quark   (Able, Abby, Brian)

This Neutron decayed via Beta minus decay caused by the nature of its W- & W+ Boson gluon fluxtube and Z Boson micro black hole structure.
3 more Quarks were formed in the Proton as the Neutron was decaying, this may have been a form of self preservation. These Quarks were:

 udu = Up Quark   (Beth,Charles,Charly)
 dud = Down Quark   (Dan, Diana, Eddie)
 udu = Up Quark   (Edna, Freddy,Frayer)
The Proton is said to carry a positive charge. The circuit is now closed.

Lepton Era/Epoch

This Epoch is said to have taken place at 1 second after the expansion of the singularity point and lasted for aprox 3 minutes after the singularity expansion. The Temperature at the begining of this process is said to have been 10^{10} Degrees Kelvin (Ten thousand million Kelvin). It cooled to 10^9 Degrees Kelvin (One Billion Kelvin) by the end of the 3 minute Era.

During the Lepton Epoch the universe was said to have been opaque.
Deutrium was thought to have been formed at around 10 seconds after the expansion if the singularity point.
Electrons and their Positrons seem to be the result of beta minus decay of the Neutron & Proton formation. In which the energy was consrved all through Proton production and some was emitted as an Electron and Positron once the Proton had become full.

Radiation at this point is said to have been between 1-10mm. The classic wavelength of the Electron or Positron is said to be 2.9×10^{-10} metres. Two Electrons are thought to be 1.22×10^{-12}m.
The mass of an Electron or Positron is said to be 0.511mev/c2
Electrons and Positrons were thought to have annihilated each other at this point with an energy of 1.022mev.

The Nucleon-Deuteron model suggests that this annihilation might have been a form of decay. Where the Electron and Positron are drawn closer together via their W- & W+ Bosons through the Z Boson. When they become as close as they can get they birth or emit a Photon and its Anti-Photon. It seems as though the Electron & its Positron might have decayed or annihilated each other while they are free flying.

They are not yet bound by the Proton because the enormously hot temperature prevents this. They are stripped away from the Proton. This might be why they decay in this state. For now the Electrons and Positrons have lost their connection to the Proton. However soon the temperatures will cool enough for the Electrons & Positrons to become slow enough that they are captured in orbit around the Proton. The Electron and Postron emit the Photon and its entangled Anti-Photon. However the Electrons energy is still conserved in the Photon /Anti Photon pair. This might happen at 1.022mev.

The mass seems to be conserved in the Anti Photon as a form of decay. While the Photon does not seem to have mass it does have energy. Since E=MC2 the mass of the Anti Photon and the energy of the Photon seem as one whole entity, particle / wave. This could be possible because the Photon & Anti-Photon should exist as close to their Z Boson as possible. They might be so close that the Anti-Photon could hold a tiny mass, that is if they do possess any mass at all. Because $E = MC^2$ we know that if something has energy it should also have an equivelent mass. However we also know that for a particle to travel as fast as light speed it should not be able to have mass. Acording to Albert Eienstien, if a particle could travel faster than light speed as a particle in the negative Anti/ Dark 'zone' should, it would not experience any time at all. Time slows down on approaching light speed. If a particle were to break the light speed barrier, the particle should not experience time. Time is a measure of change. Change is decay.

The particle in this environment should no longer decay. It should also be able to have mass because it does not know it is faster than light speed being timeless, it could have mass because the rules that say a Photon traveling at the speed of light cannot have mass should not apply if the particle does not know its speed. Therfore it should be exempt from this rule. Therefore it is possible that the Anti-Photon could carry the mass of the pair which equils the pairs energy.

Summery Of The Equations For Lepton Era/Epoch

Equations for the Lepton epoch :
C= speed of light 2.99792458×10^8 (metres per sec)
MSS = Majorana singularity speed/ velocity (metres per sec)
$\quad\quad 2.9979 \times 10^8$ m/sec
MNPS= Majorana neurtino pair speed 2.9979×10^8 m/sec
$\sqrt{Gh}$= plank time $<1.35 \times 10^{-43}$ (seconds)
pl = plank length 1.6000×10^{-35} (mm)
h= planks constant
T= temperature (Degrees Kelvin)
t= time (seconds)
λ= wavelentgh (mm)
f= frequencey (HZ)
E= energy (Joules or Gigga electron volts)
v= velocity (metres per sec)

To find the wavelength through-out the lepton era:

At 1.8125×10^{10}k the wavelength was 1.6000×10^{-11} mm
At 1.8125×10^9k the wavelength was 1.6000×10^{-12} mm

To find the frequencey at the end of the Lepton era:
$$F = \frac{MSS\ 2.9979 \times 10^8}{\lambda\ 1.6000 \times 10^{-12}\text{mm}}$$
$$\text{Frequencey} = 1.8737 \times 10^{20} \text{HZ}$$

Now we have the frequencey of the state of the universe at the end of the Lepton era:. From this we can work out its velocity by multiplying its wavelength by its frequencey.

V= $\lambda\ 1.6000 \times 10^{-12}$ mm x f1.8737×10^{20} HZ
$\quad$ velocity = 2.9979×10^8 m/per/sec
Again this shows that the velocity remained constant.

To find the Energy at the end of the period, we can use the velocity or Majorana Neutrino pair speed (MNPS):
$$E = \frac{h \times MNPS\ 2.9979 \times 10^8}{\lambda\ 1.6000 \times 10^{-12}\text{mm}}$$
Energy = 1.2415×10^{-13} J

$\quad\quad$ or

$E = h \times f\, 1.8737 \times 10^{20} HZ$
Energy = $1.2415 \times 10^{-13} J$

To find the mass we should use the Majorana Neutrino pair speed.

Mass = $\dfrac{\text{Energy } 1.2415 \times 10^{-13} J}{\text{MSS } 2.9979 \times 10^{8^2} \text{ m/sec}}$

Mass = $1.3814 \times 10^{-30} kg$

Because $E = MC^2$

$E = 1.3814 \times 10^{-30} kg \times 2.9979 \times 10^{8^2}$ m/sec

$E = 1.2415 \times 10^{-13} J$

Energy per mole = E $1.2415 \times 10^{-13} J$ x Avagadros constant AV
= $7.4765 \times 10^{10} J$

Nucleosynthesis

This Epoch began at 3 Minutes after the expansion of the singularity. It lasted for aprox 17 minutes. This brings us to 20 minutes after the expansion of the singularity point.

The temperature during this process was 10^9 Degress Kelvin at the beginning and cooled to 10^7 Degrees Kelvin by the end. Here Hydrogen was turning into Helium. This is known as the Proton Proton chain.

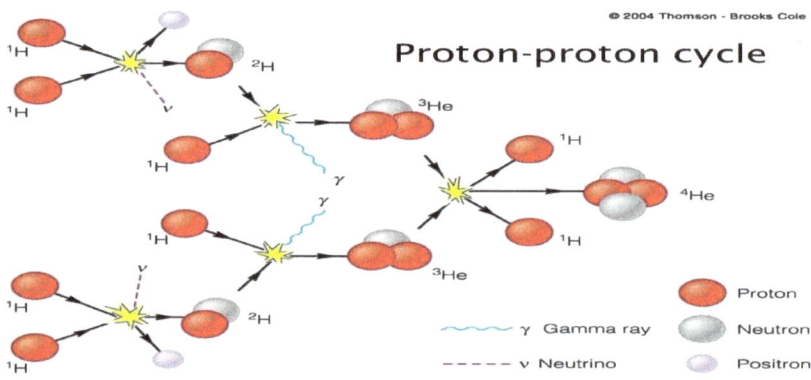

The wavelength is said to have been 2.9×10^{-10} metres by the end of the epoch. Which is less than a nanometre.

Traditional cosmology says that at this time there were 6 Protons for every Neutron. This might have been because Neutrons decay into Protons and at this point a lot of Protons have been formed. The rate of Neutron production might have been slowing down due to the temperature cooling. The Neutrons that did exist at this point in time were almost all contained inside the nuclei of Helium 4.

By the end of this epoch the universe contained 75% Hydrogeon, 25% Helium-4, 0.01% Deuterium and 0.01% Helium-3.

Summery Of The Equations For The Nucleosynthesis Era/Epoch

Equations for the Nucleosynthesis epoch :
C= speed of light 2.99792458×10^8 (metres per sec)
MSS = Majorana singularity speed/ velocity (metres per sec)
   2.9979×10^8 m/sec
MNPS= Majorana neurtino pair speed 2.9979×10^8 m/sec
$\sqrt{Gh}$= plank time $<1.35 \times 10^{-43}$ (seconds)
pl = plank length 1.6000×10^{-35} (mm)
h= planks constant
T= temperature (Degrees Kelvin)
t= time (seconds)
λ= wavelength (mm)
f= frequencey (HZ)
E= energy (Joules or Gigga electron volts)
v= velocity (metres per sec)

To find the wavelength through-out the Nucleosynthesis epoch:

At 1.8125×10^9k the wavelength was 1.6000×10^{-12}mm
At 1.8125×10^8k the wavelength was 1.6000×10^{-11}mm
At 1.8125×10^7k the wavelength was 1.6000×10^{-10}mm

To find the frequencey at the end of the Nucleosynthesis epoch:

$$F = \frac{MSS\ 2.9979 \times 10^8}{\lambda\ 1.6000 \times 10^{-10} mm}$$

Frequencey = 1.8737×10^{18}HZ

Now we have the frequencey of the state of the universe at the end of the epoch: From this we can work out its velocity by multiplying its wavelength by its frequencey.

V= $\lambda\ 1.6000 \times 10^{-10}$mm x f$1.8737 \times 10^{18}$HZ
   Velocity = 2.9979×10^8 m/per/sec
Again this shows that the velocity remained constant.

To find the Energy at the end of the period, we can use the velocity or Majorana Neutrino pair speed (MNPS):

$$E = \frac{h \times MNPS\ 2.9979 \times 10^8}{\lambda\ 1.6000 \times 10^{-10} mm}$$

Energy = 1.2415×10^{-15} J
   or

E= h x f1.8737×10^{18} HZ
Energy = 1.2415×10^{-15} J

To find the mass we should use the Majorana Neutrino pair speed.

$$\text{Mass} = \frac{\text{Energy } 1.2415 \times 10^{-15} \text{ J}}{\text{MSS } 2.9979 \times 10^{8^2} \text{ m/sec}}$$
Mass = 1.3814×10^{-32} kg

Because E=MC²

E= 1.3814×10^{-32} kg x $2.9979 \times 10^{8^2}$ m/sec
E = 1.2415×10^{-15} J

Energy per mole= E 1.2415×10^{-15} J x Avagadros constant NA
E = 7.4765×10^{8} J

Matter Epoch & Secondary Inflation Era/Epoch

This Era began at 17 minutes after the expansion of the singularity and lasted until 380,000 yrs. The temperature at the begining of this Epoch was 10^7 degrees Kelvin and cooled to 3000k by the end of the era. Cosmology says that by this time the universe was like a giant star.

After Nucleosynthesis where the star-like universe had been burning Hydrogeon into Helium, the star might have burnt its fuel which should have caused it to expand into a Red Giant. This should have happened between Nucleosynthesis & Matter epoch. This expansion into a Red Giant should have cause a secondary inflation period.

During the Matter epoch more elements should begin to form at the core of the universe although surface temperatures should have been much cooler.

Summery Of The Equations For Matter & Secondary Inflation Era/Epoch

Equations for the Matter epoch :
C= speed of light 2.99792458×10^8 (metres per sec)
MSS = Majorana singularity speed/ velocity (metres per sec) 2.9979×10^8 m/sec
MNPS= Majorana Neurtino pair speed 2.9979×10^8 m/sec
$\sqrt{Gh}$= plank time $<1.35 \times 10^{-43}$ (seconds)
pl = plank length 1.6000×10^{-35} (mm)
h= planks constant
T= temperature (Degrees Kelvin)
t= time (seconds)
λ= wavelength (mm)
f= frequencey (HZ)
E= energy (Joules or Gigga electron volts)
v= velocity (metres per sec)

At 1.8125×10^7k the wavelength was 1.6000×10^{-10} mm

To find the frequencey at the end of the Matter epoch:

$$F = \frac{MSS\ 2.9979 \times 10^8}{\lambda\ 1.6000 \times 10^{-10} mm}$$
$$\text{Frequencey} = 1.8737 \times 10^{18} HZ$$

Now we have the frequencey of the state of the universe at the end of matter era:. From this we can work out its velocity by multiplying its wavelength by its frequencey.

V= $\lambda\ 1.6000 \times 10^{-10}$ mm x f1.8737×10^{18} HZ
   Velocity = 2.9979×10^8 m/per/sec
Again this shows that the velocity remained constant.

To find the Energy at the end of the period, we can use the velocity or Majorana Neutrino pair speed (MNPS):
E= $\frac{h\ x\ MNPS\ 2.9979 \times 10^8}{\lambda\ 1.6000 \times 10^{-10} mm}$
Energy = 1.2415×10^{-15} J
   or

$E = h \times f\, 1.8737 \times 10^{18} HZ$
Energy = $1.2415 \times 10^{-15} J$

To find the mass we should use the Majorana Neutrino pair speed.

$$\text{Mass} = \frac{\text{Energy } 1.2415 \times 10^{-15} J}{\text{MSS } 2.9979 \times 10^{8^2} \text{ m/sec}}$$
$$\text{Mass} = 1.3814 \times 10^{-32} kg$$

Because $E = MC^2$
$$E = 1.3814 \times 10^{-32} kg \times 2.9979 \times 10^{8^2} \text{ m/sec}$$
$$E = 1.2415 \times 10^{-15} J$$

Energy per mole = $E = 1.2415 \times 10^{-15} J \times$ Avagadro's constant NA
$$= 7.4765 \times 10^{8} J$$

De-coupling Era/Epoch

The De-coupling Epoch began at 380,000yrs years after the expansion of the singularity point. The Epoch began with a Temperature of 10^7 Degrees Kelvin and ended at 3000 Degrees Kelvin.
The energy created at this time would have been 13.6ev of Ionisation energy.

Cosmology says that at the beginning of this epoch the temperature was too hot for Electrons to be in orbit around Protons.

The surface of the star-like universe should have been much cooler than the core.

At the begining of this epoch, Electrons had been stripped from their Protons due to high temperatures. It takes 13 Gev of ionisation energy to strip Electrons from Protons. The Electrons bounce off other free Electrons. The result is a an opaque soupy looking star like formation.
Photons at this stage would not have been able to travel out to the edges of the star-like universe because they would have been bumping into the free flying Electrons and Protons.

By the end of the epoch temperatures had fallen enough for Electrons to drop into orbit around ionised Hydrogen and Helium Nuclei. This caused a drop in free electron density. With less density the universe became transparent. Photons were able to move more freely. Compton scattering of Photons was aprox equil to the rate of the expanding universe.

It takes from between 150,000 to 200,000yrs for a Photon of light to escape from the core of this hot gassy star like universe. It has to first pass through the radiation zone then through the convection zone.

This is the point at which the CMB can be detected. The CMB picks up a signal of the earliest light of the universe. This radiation is uniform in all directions. This tells cosmologists that at one time the universe must have been very small and grew rapidly.

As we know the universe at this point was like a super-mega star. Light travels in all available directions at a constant speed. Therfore the light that was emmited from this star would have been uniform in all directions. We know that large stars eventually experience supernova explosions, in which their elements are projected outward to form nebulas. The nebula of such a star after its first supernova might be equil to the size of a solar system which is the size that the CMB reports. However the CMB reports two such events where it detects two distinct bands of radiation.

The nebula goes on to form galaxies & planets. Therefore it is quite possible that the inflationary period predicted by the CMB was a result of the early star like universe which had just began to release radiation. The bands detected in the CMB might have been it's supernova explosions.

Summery Of The Equations For The De-Coupling Era/Epoch

Equations for the Decoupling epoch :
C= speed of light 2.99792458×10^8 (metres per sec)
MSS = Majorana singularity speed/ velocity (metres per sec)
    2.9979×10^8 m/sec
MNPS= Majorana Neurtino pair speed 2.9979×10^8 m/sec
$\sqrt{Gh}$ = plank time $< 1.35 \times 10^{-43}$ (seconds)
pl = plank length 1.6000×10^{-35} (mm)
h= planks constant
T= temperature (Degrees Kelvin)
t= time (seconds)
λ = wavelentgh (mm)
f= frequencey (HZ)
E= energy (Joules or Gigga electron volts)
v= velocity (metres per sec)

At 3000 k the wavelength was 9.6667×10^{-7} mm

To find the frequencey at the end of the De-coupling epoch:

$$F = \frac{MSS\ 2.9979 \times 10^8}{\lambda\ 9.6667 \times 10^{-7} mm}$$

Frequencey = 3.1014×10^{14} HZ

Now we have the frequencey of the state of the universe at the end of the de-coupling epoch:. From this we can work out its velocity by multiplying its wavelength by its frequencey.
V= $\lambda\ 9.6667 \times 10^{-7}$ mm

h x f 3.1014×10^{14} HZ
   Velocity = 2.9979×10^8 m/per/sec
Again this shows that the velocity remained constant.

To find the Energy at the end of the period, we can use the velocity or Majorana Neutrino pair speed (MNPS):

E= $\frac{h \times MNPS\ 2.9979 \times 10^8}{\lambda\ 9.6667 \times 10^{-7} mm}$
Energy = 2.0550×10^{-19} J
          or

$E = h \times 3.1014 \times 10^{14}$ HZ
Energy = 2.0550×10^{-19} J

To find the mass we should use the Majorana Neutrino pair speed.

$$\text{Mass} = \frac{\text{Energy } 2.0550 \times 10^{-19} \text{ J}}{\text{MSS } 2.9979 \times 10^{8^2} \text{ m/sec}}$$
$$\text{Mass} = 2.2865 \times 10^{-36} \text{ kg}$$

Because $E = MC^2$
$$E = 2.2865 \times 10^{-36} \text{ kg} \times 2.9979 \times 10^{8^2} \text{ m/sec}$$
$$= 2.0550 \times 10^{-19} \text{ J}$$

Energy per mole = E 2.0550×10^{-19} J x Avagadro's constant NA
$$= 1.2376 \times 10^{5} \text{ J}$$

Radiation Era/Epoch

This epoch is known as the Dark ages of the universe. It began at the moment of De-coupling.

The Epoch lasted for aprox 150 Million years.

The Temperature was still at 3000 Degrees Kelvin at the beggining of this Era. The wavelength was 970mm.

$$\Lambda = \frac{0.0029}{3000k} = 970nm$$

970 Nano metres is in the range of Infra-red radiation. This means that the universe was not radiating visible light at this time, it would have still been dark.
Energy levels were low.

Summery Of Equations For Radiation Era/Epoch

Equations for the Radiation epoch :
C= speed of light 2.99792458x10^8 (metres per sec)
MSS = Majorana singularity speed/ velocity (metres per sec)
    2.9979x10^8m/sec
MNPS= Majorana Neutrino pair speed 2.9979x10^8m/sec
√Gh= plank time <1.35x10^-43(seconds)
pl = plank length 1.6000x10^-35(mm)
h= planks constant
T= temperature (Degrees Kelvin)
t= time (seconds)
λ= wavelength (mm)
f= frequencey (HZ)
E= energy (Joules or Gigga electron volts)
v= velocity (metres per sec)

The wavelength during Radiation epoch was 3000K. The wavelength was 9.6667x10^7mm.

To find the frequencey at the end of Radiation epoch:

$$F = \frac{2.9979 \times 10^8 \text{m/sec}}{9.6667 \times 10^{-7} \text{mm}}$$
$$F = 3.1013 \times 10^{14} HZ$$

Now we have the frequency of the state of the universe at end of the era. From this we can work out its velocity by multiplying its wavelength by its frequencey.

V= λ 9.6667x10^-7mm x 3.1013x10^14HZ
   velocity = 2.9979x10^8 m/per/sec

Again this shows that the velocity remained constant.

To find the Energy at the end of the period, we can use the velocity or Majorana Neutrino pair speed (MNPS):

$$E = \frac{h \times MNPS \; 2.9979 \times 10^8}{\lambda \; 9.6667 \times 10^{-7} \text{mm}}$$
Energy = 2.0550x10^-19J
           or

$$E = h \times 3.1013 \times 10^{14} HZ$$
$$\text{Energy} = 2.0550 \times 10^{-19} J$$

To find the mass we should use the Majorana Neutrino pair speed.

$$\text{Mass} = \frac{\text{Energy } 2.0550 \times 10^{-19} J}{\text{MSS } 2.9979 \times 10^{8^2} \text{ m/sec}}$$
$$\text{Mass} = 2.2865 \times 10^{-36} kg$$

Because $E = MC^2$
$$E = 2.2865 \times 10^{-36} kg \times 2.9979 \times 10^{8^2} \text{ m/sec}$$
$$E = 2.0550 \times 10^{-19} J$$

Energy per mole = E $2.0550 \times 10^{-19} J$ x Avagadro's constant NA
$$E = 1.2376 \times 10^{5} J$$

Re-Ionisation Era/Epoch

This Epoch lasted for 150million years.

The Re-Ionisation Energy was 13.6ev for Hydrogen.
The wavelength was 91.3 nanometres.

Cosmology says that at this point new stars beging to form. Containing more elements. These new stars radiate Ultra Violet radiation. This new radiation of 91.3 nano metres became Re-ionisation energy. It caused Electrons to drop back into orbit around Protons and allowed stable atoms to form.

The wavelength here was thought to have been 91.3 nanometres. Here stars radiated with ultra violet radiation. This UV radiation causes Ionisation to occur again stripping Electrons from their nuclei a second time. The universe was opaque again.

The Equation is:
$$\lambda = \frac{R(6.626 \times 10^{-34} sec) c(3 \times 10^8 m/sec)}{E(13 Gev)(16 \times 10^{19} J/ev)}$$
$$= 91.3 \text{ nanometres of UV radiation.}$$

For 150-1 billion years atoms become so far apart that they do not interact with radiation as much. This causes the universe to be transparent.

Between 300-500 million years on, Gravity caused pockets of Primordial gas to become more dense. The pockets collapsed under their own gravity. This caused them to become hot enough to trigger nuclear fusion between the Hydrogen atoms. The first supermassive stars were created.

Next the first Galaxies form. This happened 480 million years after the expansion of the singularity point.

Supermassive Star-like Universe

Hydrogen atoms which contained one Proton would have fused together creating Deuterium. Next Deuterium atoms which contain one Neutron and one Proton would have fused together, creating Helium. This should have happened during Nucleosynthesis epoch, which lasted 17 minutes. The star-like universe at the begining of Nucleosynthesis should have been 10,000km big and expanding through-out the epoch.

During the process the core would have been around 15 million degrees while the surfce should have been around 6000 degrees. The process of burning Hydrogen into Helium stops when the Hydrogen in the star's core runs out.

At this point energy should have been released creating pressure which caused the super-massive star like universe to expand. It should have become similar to a Red Giant. This should have cause a secondary inflamatory period.

This phase should have happend between Nucleosynthesis and Matter era. Where the temperature was hot enough for this to happen at around 100 million degrees kelvin.

Red Giants collapse in on their selves during the end of their life. As the star collapses in on itself its atomic structure would have been disturbed. This could have resulted in Electrons at the star-like structures surface, being stripped from their Protons. If this happened no light would be visible.

As the star-like structure is collapsing in on itself pressures at its core cause Helium atoms to fuse together, the temperature should have been around 100 milion degress at the core. Helium is burnt to form Carbon and Oxygen.
With each stage the core should collapse furthur. We know that at this point when a star has burnt all of its fuel fusion stops. No more energy can be released.

The star-like structure should fold in on itself then re-bound outward. As the star-like universe re-bounds outward. De-Coupling era is ending. This results in Electrons and Protons no longer being crushed and disturbed, the Electrons should drop into orbit around Protons. As the star-like universe re-bounds, more space becomes available for Photons to move without collisions and escape to the surface. This event should be observable in the CMB radiation as the first band where the first Photon radiaton is released.

From the time the star re-bounds Radiation epoch should have taken place. During the epoch the radiation we should see is Infrar red radiation. The universe should have once again been plunged into darkness. By the end of the era Neutrinos would have finally reached the surface of the super-massive star-like universe. When this happened at the end of the Radiation epoch, the star should have exploded releasing its elements out as a cloud nebula.
This event should also be visible in the CMB signal as the second band.
Gravity causes the atoms to congregate into reigions. Gravity causes pockets of primordial gas to become dense enough that they collapse under their own weight. This should have caused a disruption of the atomic stuctures of the particles. The event might have caused Electrons to become stripped from their nucli for a short time. Temperatures would have reached a point where nuclear fusion could take place between the Hydrogeon atoms and new supermassive stars were born.

The new stars cause UV Re-ionisation energy which results in any stray Protons & Electrons who have been disturbed to become re-ionised and stable.

The Cosmic Neutrino Background radiation is thought to detect Neutrino de-coupling. The event happened within the first second of the universe lifespan. It is thought to have been a cosequence of weak interactions between neutrinos. This could have happened durng the time of the initial inflamatory period where Primordial Majorana Neutrinos were reproducing at a fast rate.

Scientists at Oxford, England have found that at this point the star released Neutrinos. Ian Chapman from Oxford is one of the Scientists who work on the Jet star machine. They have found that when the core colapses 99% of the energy released is in the form of Neutrinos.
They have detected Neutrino bursts which escape into the cosmos at almost light speed, just a few hours before the star is seen to go super-nova. They say that the Neutrinos make up the blast waves which ripple up through the star and collide with the stars outer layer. When they break through the outer layer the star explodes. Scientists at Oxford say that the Neutrinos reach us a few hours before the emitted light from the super-nova. Therfore we now have a way to detect when a star is about to die.

We know that the early universe was like a giant star. We know that all stars are born and die. The early universe's star-like structure should have been no different. We know that its star-like structure can not still exist or we would be living inside a giant star. Its energy must have dispersed amoung the universe. The only way a star does this is by dying and in the process creating more stars. Therfore our early super-massive star-like universe must have experienced a super-nova environment. This explosive event at the end of its life would have been the "big Bang". Before it went supernova it lived. During the early universes super-massive star-like state it birthed space-Time itself.

Next we should see young stars begining to develop in this cloud Nebula, stella nursery. These stars have life cycles of their own and also die, creating new stars in the nebula as they do.
The heavier elements in the nebula clump together creating clumps of mass. The lighter elements in the nebula move toward the outer regions of the masses. The heavier elements congregate toward the central regions. These clumps interact and react with each other, harden and Eventualy form planets. The planets form solar systems which in turn form galaxies in our universe. Black holes are usually detected at the center of galaxies. The black holes might have been formed when the star goes supernova. The star on a subatomic level is made from Neutrinos of different atoms. It is possible that the star has a similar structure as the Neutrinos which has developed from particle programming. If it were similar it might also have poles and a vaccum (0) point at its very core.

When the star explodes in a supernova the centre might still be trying to pull Everything back together. This would lead to the formation of a black hole at the center of what will become a galaxy. The black hole continues over billions of years to pull the debris toward itself, devouring the galaxy a little at a time over the ages. While the black hole is busy pulling debris into itself some energy escapes on contact with the Event horizion. The energy seen here is emmitted in the form of gamma rays which also emit Neutrinos. The Neutrinos spew out in gamma ray jets far into the cosmos ready to form new Quarks which become atoms for new galaxies. This proccess would continue until the black hole had consumed all of its galaxy and the original whole galaxy system would enter a state of entropy. The new Neutrinos would continue to create a new galaxy in its place. This galaxy would begin in a similar way to how the first grew. The Neutrinos form Quarks and more Hydrogen is formed, next Deuterium is formed, a star is born, goes supernova and so on in an endless oroborus cycle.

If Light Is The Fastest Thing In The Known Universe, How Did The Dark Get There First?

The Electron was discovered by Sir Joseph John Thomson in 1897. He also found isotopes of different elements. He was awarded the 1906 Nobel Prize for Physics.

The Electron / Positron entangled pair are now orbiting the nucleus and try to break free. However they are still caught up via entanglement between their own W Boson gluon flux tube body and the Z Boson 0 point/ micro black hole. The Electron is said to carry a negative charge that is opposite to the positive charge of the Proton. The Electron and Positron are also attracted to the Z Bosons /micro black holes inside the flux tubes, belonging to the Neutrinos inside the Proton structure. Like celestial black holes, the Z Bosons should cause a Gravitational pull strong enough to keep the Electron & Positron in orbit as long as they remain close enough. This is Gravity working in its tiniest form. The Gravitational pull might be interpreted as 'Micro Gravity'.

The Electron & Positron pair are still entangled with each other as they orbit the Nucleon in shells whilst gaining speed & energy. The particle that was once (+1) Georgina has turned into a (-) Positron.

The Positron would now carry the information and therefore the mass. It exists below (0) state and is considered Anti or "Dark" as (-1). This Dark energy of the Positron might be traveling faster than its Electron counterpart because it is located in the below (-1) state which should, according to the laws of relativity be an environment where particles can move faster than light speed. Therefore it is possible that the Electron is dragging the Positron around with it. As they become faster and move up into the outer orbit of the atom the Positron drags the Electron closer to light speed. This is because of the nature of their entangled gluon flux tube /W-& W- Boson body that holds them together. The head Electron and tail Positron are being attracted to each other. The Positron pulls the Electron towards itself and ultimately toward its own body centre (0) point.

During nuclear fusion the Electrons and Positrons of heavier atoms reach a state where the Electron is on the "event horizion" of the Z Boson (0) transition Alpha decay zone. Here the Electron should partly decay. This means that it should lose some of its energy. This energy is emitted as a Photon. The Positron partly decays into an Anti-Photon. After the Electron and Positron emit the Photon and Anti-Photon energy, it drops back down in energy state. The creation energy has traveled through the Neutron into the

Proton then out into the cosmos. The Neutrons, Protons and Majorana

Neutrinos do not fully decay but lose energy in this process. They might later regain this energy when more creative energy passes though them. The atom is constantly creating more energy which allows them to survive for so long. The atom is like a tiny energy factory. When the energy is finally free from the Nucleus in the form of a Photon and Anti-photon it is able to travel through the universe for years.

Photon Mass = 0
Travels at 300,000,000 meters per sec which is equivelent to 186,000 miles per sec.

Photons have an extreamely long life. They seem to hover right on the Event horizon of the Z Boson/ micro Black hole structure of its own entangled body. It stays dangerously close to its own tail end which might carry its information and mass. Hypothetically, if the Photon were Ever to decay through the (0) point it would become an Anti-particle. It would be united with its tail end to become an Anti-Photon / Anti-light pair. Scientists have suggested that Photons might have a very small mass. If this is true then the laws of physics would allow the Photon to eventually
decay. Some suggest that the Photon might decay back into Neutrinos. However the Nucleon-Deuteron model suggests that it would decay straight into the (-) dimension. No-one has Ever witnesed the decay of a Photon. The light that was emitted at the begining of the universe can still be detected, this means that Photon's minimum life span is at least 1 billion, billion years.

The Anti-Photon must have the ability to travel faster than light or Photon speed, because it exists in the below (0) state which should, according to the laws of relativity be an environment where particles can move faster than light speed. Therefore the Photon must be led around by its Anti-Photon tail end which carries the information and mass. This could be why we find it difficult to detect Dark Matter and Dark Energy. It travels faster than light and exists in a dimension of anti or "decayed" particles. These decayed particles only appear decayed. They mearly exist within a different state, with opposite properties. They still make up the mass of our universe and can effect the gravity of whole galaxies. They make up the "missing" mass of our atoms but they are not "missing". They are very real and very functional.

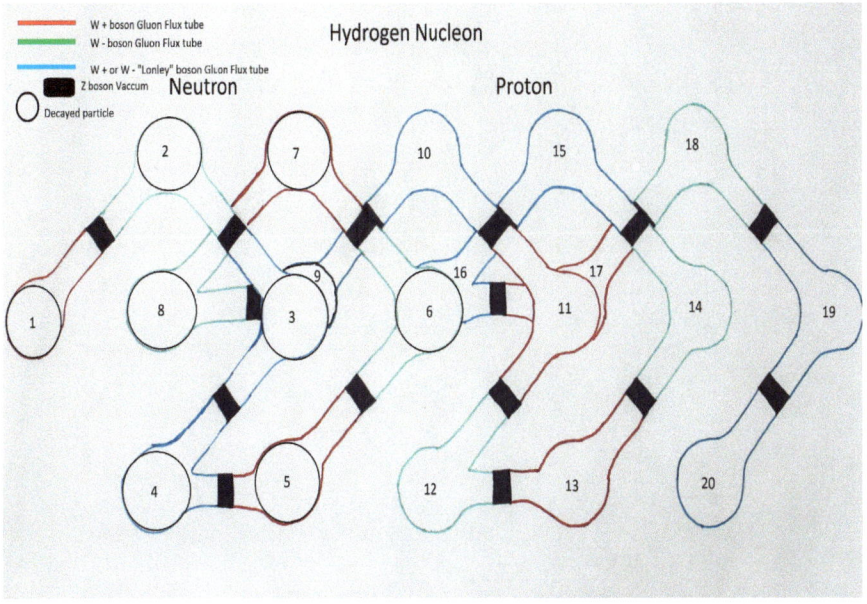

Hydrogen with decayed Neutron.

124
Hydrogen with decayed Neutron and Higgs Boson adding stability.

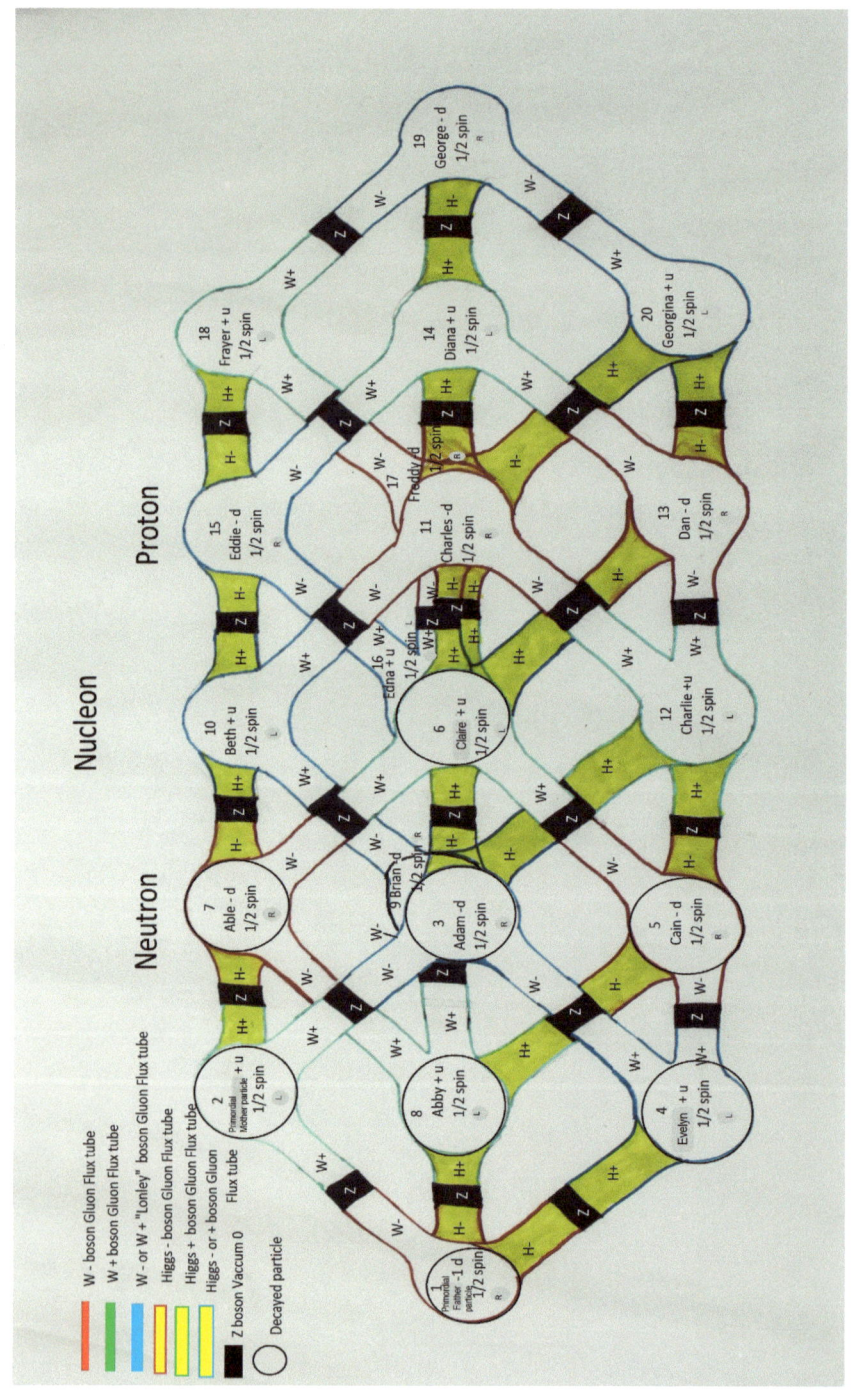

DEUTERIUM (AKA) Heavy Hydrogen
In The Nucleon-Deuteron Model

We know that in the sun it takes Hydrogen around 1 Billion years for it to make the evolutionary step to become Deuterium (H2) This would have been the case with the very first Hydrogen atoms that were developed at the "time" or state of the begining of the universe.

It is thought that it took a long time for the number of Hydrogen atoms to grow to amounts that they could collide. The collision of Hydrogen atoms causes them to fuse together. The Electrons from the atoms attract to the Proton of the other. Each Hydrogen atom has 1 Proton and one Electron and a Positron. Together the two Hydrogen atoms create an atom that has 2 Protons and 2 Electrons and 2 Positrons. Protons are usually repeled from each other, it is nuclear Weak force that prevents the Protons of each atom from repelling each other in atomic structures. The two Protons fuse forming a Di-Proton. The Di-Proton experiences Beta plus decay and one Proton changes its state into a Neutron. Simultaneously the extra Electron and Positron are released as a Photon and Anti-Photon pair.

The Proton which reverts back into a Neutron state, might inhabit the same physical (+1) energy space where the original Neutron used to be before it decayed.
This process would balance out the atom and create a stable foundation on which heavier more complex atoms can be built. This seems to be another self preservation or repair technique, where the Nucleon sends a Proton to do the job of a decayed Neutron by replacing it. One of the two Electrons and a Positron are released as a Photon and Anti-Photon. The new atom containing 1 Neutron, 1 Proton, 1 Electron and a Positron is called Deuterium or "Heavy Hydrogen".

Full Deuterium with Higgs mechanism

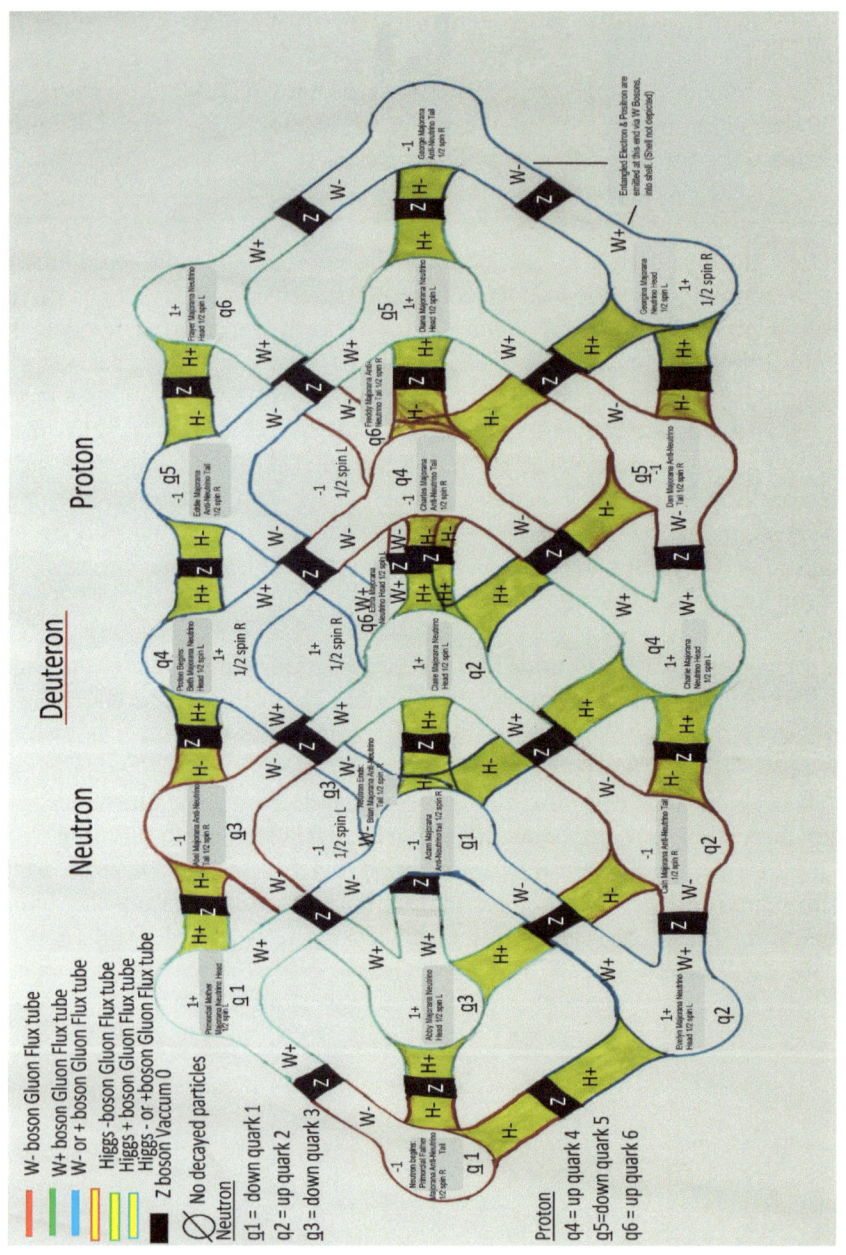

Results

Tables show Majorana Neutrino & Anti-Neutrino qualities during Hadron formation.

Quark 1

Name	Primordial Father	Primordial Mother	Adam
State	-d	+u	-d
Spin	1/2	1/2	1/2
Paired/Lonley	p	p	L
Colour	Red	Green	Blue
Boson	W-	W+	W-
Binary	0	1	0

Quark 1

Quark Flavour	q down
Approx Mass	4.8 mev/c2
Spin	1/2
Electric charge	1/3

Quark 2

Name	Evelyn	Cain	Claire
State	+u	-d	+u
Spin	1/2	1/2	1/2
Paired/Lonley	L	P	P
Colour	Blue	Red	Green
Boson	W+	W-	W+
Binary	1	0	1

Quark 2

Quark Flavour	q up
Approx Mass	2.3 mev/c2
Spin	1/2
Electric charge	2/3

Quark 3

Name	Able	Abby	Brian
State	-d	+u	-d
Spin	1/2	1/2	1/2
Paired/Lonley	P	P	L
Colour	Red	Green	Blue
Boson	W-	W+	W-
Binary	0	1	0

Quark 3

Quark Flavour	q down
Approx Mass	4.8 mev/c2
Spin	1/2
Electric charge	1/3

Quark 4

Name	Beth	Charles	Charlie
State	+u	-d	+u
Spin	1/2	1/2	1/2
Paired/Lonley	L	p	p
Colour	Blue	Red	Green
Boson	W+	W-	W+
Binary	1	0	1

Quark 4

Quark Flavour	q up
Approx Mass	2.3 mev/c2
Spin	1/2
Electric charge	2/3

Quark 5

Name	Dan	Diana	Eddie
State	-d	+u	-d
Spin	1/2	1/2	1/2
Paired/ Lonley	P	p	L
Colour	Red	Green	Blue
Boson	W-	W+	W-
Binary	0	1	0

Quark 5

Quark Flavour	q down
Approx Mass	4.8 mev/c2
Spin	1/2
Electric charge	1/3

Quark 6

Name	Edna	Freddy	Frayer
State	+u	-d	+u
Spin	1/2	1/2	1/2
Paired/Lonley	L	P	P
Colour	Blue	Red	Green
Boson	W+	W-	W+
Binary	1	0	1

Quark 6

Quark Flavour	q up
Approx Mass	2.3 mev/c2
Spin	1/2
Electric charge	2/3

Name	George	Georgina
State	-d	+u
Spin	1/2	1/2
Paired/Lonley	p	p
Colour	Blue	Blue
Boson	W-	W+
Binary	0	1

creates

Name	Electron	Positron
Approx Mass	0.511 mev/c2	
Spin	1/2	

The patterns found in the Nulcleon are as follows:

Paired or lonley	PPL	LPP	PPL	LPP	PPL	LPP	L&P
colour	RGB BRG		RGB BRG		RGB BRG		BB
Binary	010	101	010	101	010	101	01

Boson pattern:
W- W+ W-,
W+ W- W+,
W- W+ W-,
W+ W- W+,
W- W+ W-,
W+ W- W+,
W- W+.

Helium 4

When Deuterium is finally formed two of these atoms fuse together and create Helium 4. This atom has 2 Neutrons, 2 Protons, 2 Electrons and 2 Positrons.

Spooky Action At A Distance

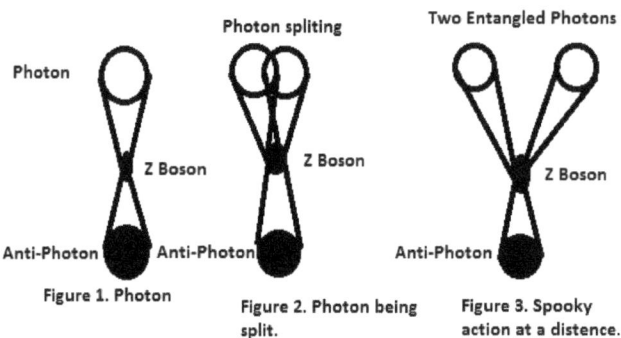

Figure 1. Photon

Figure 2. Photon being split.

Figure 3. Spooky action at a distence.

When a Photon beam is split, the Photons remain entangled and know the state of the other Even at great distances.

The Nucleon-Deuteron model suggests that the cause of this might be due to the Anti-Photons envolvment with each other. When we split a Photon we split its (+) half. However its Anti-particle might still remain unsplit. This means that the two parts of the split Photon are still in communication with its whole and united Anti-self. If we remember that the Flux- tube W Bosons retain their density even at a distance, we can see how the Photons could stay in contact even at a distance. The result is that split Photn half 1 is linked to the unsplit Anti-Photon and through this connection knows instantly what Photon half is doing because Photon 2 is also in communication with the Anti-Photon. The Anti-Photon is capable of talking to both halfs of the split Photon simultaniously. If we were to also split the Anti-Photon we might find that the two halfs of the Photon were no longer able to communicate at long distances because they have had their connection severed.

"Innate" Particle "Memory"

The model shows that once a particle has been though the evolutionary process, it retains "innate memory" or blueprint of all of its previous states to be called upon at a later "time". This would happen if a change to the particles environment was unfavourable and the particle needed to revert to a previous state for self preservation.

This is similar to how a human baby grows inside of the womb. A dormant egg containing a copy of 23 genes inherited form the mother can be compared to the dormant Neutrino head (+1) which contains information inherited from its (+1) mother Neutrino head. Inside the womb the egg is activated by the active sperm which also contains 23 genes. Likewise the Neutrino head inside the Nucleon womb like environment is entangled with a (-1) Neutrino active energy carring tail end. It carries more energy because it has more mass due to it being an Anti- particle which exists beyond the (0) midpoint.
Together the egg & sperm produce a cell cluster which evolves through all of the evolutionary states that hunan beings have lived and evolved through over the ages from creation to now.

Science has showed how the developing fetus has fish-like gills which disappear or evolve into the ear structure. We have a tail which is absorbed by the body. After just 9 months we have gone through all of our entire human evolutionary process inside our mothers womb.

Our DNA remembers all of our evolutionary steps in case we should one day need to call upon them. The information contained in our DNA seems to derive from the atoms that make up the Amino acids in our protines. The Amino acid also has a Carboxoyl group or side chain which dictate what type of Amino acid the final structure will become. The Amino Acids join together because atoms from the side chain attract together forming peptide chains.

When peptide chains join they form a protine. These protines fold upon themselves due to the Atoms in the side chains attracting to each other. Therfore the types of atoms in the side chains directly effects the way the atoms attract and the pattern or code for protine folding will differ. Because each type of protine is folded differently, each type is useful for different purposes. Amino acids are grouped together in threes called codons. Each group of three is connected to another and so on. Each Amino acid corresponds to a letter. Groups of three Amino acids have three letters. These letters are the genetic code.

Amino acids are composed of Carbon, Hydrogen, nitrogen and oxygen. Side chains contain various other atoms.

We can see that Hydrogen and all other atoms after carry information, that they must have inherited from the Majorana singularity particle. With each stage of their evolution the particles changed in order to survive or as a means of self preservation. They retained this information. As the atoms form Amino acids they still have this inherited information or code. As the Amino acids join they each learn new information which passes on through particle generations. The information gets inherited by the protine as a whole and into the DNA coding. As DNA forms from protines it too adapts and learns, it passes on the code through replication.

The most primitive particles had a desire for self preservation. This led to attraction and energy exchanges. These interactions were benificial to the particles survival, so the particle learns this and evolves ways to better survive. These survival techniques include bonding with other particles to become stronger. Exchanging energy and having the ability to 'sense' what it needs for survival. As Amino acids form into protines then into cells, the structure uses the information to build upon. Where there is a need to survive, the lifeform knows about it on an atomic level and its side chains are adjusted accoringly so that the cell can produce a different function. Evolution of spiecies must have developed due to altering the side chains of Amino acids to produce the required coding which would produce the protine needed for sencery perception and survival. Through this method Single cells evolve into multicellular organisms.

First Particle Oscilations In The Universe

Energy moves from (-) to (+) then back into (-) due to the replicative nature of the Neutrino. When this happens the process appears to cause a primitive first oscillation process. When the diagram is viewd it appears that the structure consists of tiny oscillations which are caused by the energy moving from a higher state to a lower state.

The energy of the up section of the oscillation is always (+) while the energy of the down section of the oscillation is always (-). This is because there is a transition zone at mid- point where the (0) zone is situated.

The energy path resembles a cube structure rather than a straight line path that we see in some other energies. The cube energy seems to create this stucture because of the Neutrinos attraction and replicative system. The energy seems to move in tiny straight lines and accommodates the process of Neutrino replication. This indicates that this is a square wave. The lines join at right angles which seem to resemble a folding pattern that is spiraled. The finnished structure of the Nucleon or Deuteronresembles a cube-like structure.

Therefore the reproductive system is the driving force for the oscillations which cause the color charge of the Quarks. When the energy emerges out of the otherside of the cube-like stucture as an Electron and Positron, it wiould still be travelling in a straight line if it was not in orbit around the nucleus. If it were to break free it would continue to travel out in a straight line. When a Photon is eventualy released it is free to travel in a straight line and it's only attraction is its own Anti-Photon. It does not form a cube structure because at this point it is not reproducing but has gone through a small state of decay. It seems to be in a state as close to total decay as is possible, existing on the brink of the event horizion of the Z Boson/ micro black hole between itself and its Anti-particle.

If we could see the oscillations or color flavors, we might detect colour flashes while the energy transitions from one Neutrino state to the next while the Quarks are forming. We know that ultimately this structure will release a Photon.
We know that if we refract a Photon beam with a prism or water droplet we get a rainbow. Light slows down when it passes through a prism. The shorter wave lengths slow down more than the longer wavelengths. The differences in the speed of the wavelengths causes the Photon beam to separate into the 7 colours that we see. These colours are made of just 3 primary colours, Red,

Green and Blue.
When they combine they form white light. However once refracted they can be seen as distinct colours. If we could see inside of the Nucleon we might be able to detect the Red, Green and Blue primary colours. As the first Primordial Father Anti-Neutrino (Red) sends its energy through to the first Primordial Mother Neutrino (Green), we might see a temporary yellow glow. This is because red and green make yellow. The Green Primordial Mother sends her energy to the Blue Adams Anti-Neutrino head which might produce a cyan glow because green and blue make cyan. When Blue Evelyn passes her energy to Red Cain we might see a temporary magenta glow because blue and red produce magenta. These are known as the secondary colours of light.
The Quark colours must move from red through green to blue because that is the order they appear in the refracted rainbow Photon beam.

It appears that oscillations are not only caused by Neutrinos flipping during flight as the Takaaki Kajita & Arthur McDonald theory suggests. The oscillations seem to also be caused by bound Neutrinos replicating because they have flipped. These waves are square waves.

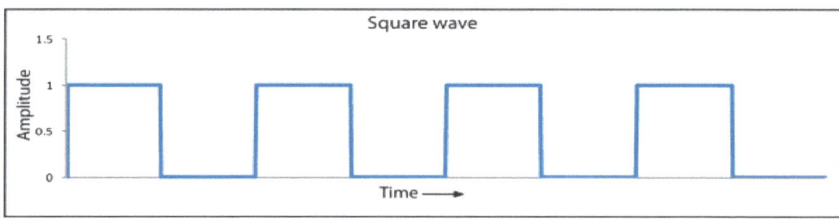

Electron Neutrinos have been found to oscillate. They seem to change from Electron Neutrino to Muon Neutrino to Tau Neutrino. Electron Neutrinos are emitted from Electrons. During flight over many kilometres they might change to Muon Neutrino to Tau Neutrino as a form of self preservation.

If Electrons are not orbiting the nucleus they might not emit Photons. If an electron were to decay it should emit its energy and an Electron Neutrino can be found. Like-wise the positron should emit its Anti-Neutrino. The Electron Neutrino and its Anti-particle should should function the same as Majorana Neutrino pairs. Where the energy of the Neutrino /Anti-Neutrino heads & tails is pulled toward the centre Z Boson. As this happens a change in state should occur due to the mass being effected.

The Neutrinos new state should be a Muon. As the particle travels it is desperately trying to pull itself together but at the same time it is trying not to collapse in on itself through it's centre Z Boson. The result is that it's state alters again to become a Tau Neutrino. The cause of it's formation into a Tau

Neutrino should be that under some circumstances the Tau Neutrino is said to have the ability to form into Hadrons. Apart from the Majorana Neutrino / Anti- Neutrino pairs this is the only one which has this inherited ability. So we can see that it appears as though the Neutrino particle is trying to preserve itself by trying to build an entire Hadron through Majorana Neutrino reproduction, so the process can begin again.

Neutrinos Don't Break The Speed Limit But Anti-Particles Might, Time Travelling Neutrinos Explained

Scientists at Opra, Italy announced the results of an experiment, which found Neutrinos travelling 60 nano seconds faster than the speed of light. Neutrinos had been sent from CERN Switzerland to Opra, Italy.
However Opra later announced that they had found an error in their equipment and it was possible that they did not travel faster than light after all.

The Nucleon-Deuteron Model suggests that Neutrinos travel close to the speed of light. It is also possible that Anti-particles might travel faster than light speed due to their state. If this were the case then a Neutrino being sent to a receiver might appear to be received before the Neutrino was sent. This is because the Anti-Neutrino tail end carries most of the information. If it were to travel slightly faster, then the Neutrino's head entangled end, the tail end would reach the receiver, carrying the information before it's head end that we can detect. So the information would get there before e knew the neutrino head end had arrived.

The tail would also have left the sender before the head was detected to have left. The receiver should pick up the Anti-Neutrinos message and send a message back before it had detected the Neutrino heads arrival.
The sender might think that they had received a message back before they sent the first one out. The effect appears to be time travelling Neutrinos.

Discussion

Black Hole V Majorana

It was concluded that the black hole seems to function as a giant recycling system. It pulls matter into itself causing it to stretch out much the same as the Majorana particle does when it is in its do-nut form. The Black holes ability to attract matter into itself might originate from its ancester the Majorana's design. The center may be a 0 vacuum, but nearer the outer edges the black hole might have a charge which is caused by stretching similar to the heat sink stretching of the Majorana particle. When the black hole stretches out to a certain point its center might become stretched out causing polarity to form. This polarity may be observed by the direction of the gamma ray emissions. The polarity causes Electro-Magnetic energy and can attract matter to itself. This matter has been observed to be spinning in a spiral shape into the black hole.

A black hole is large, it has a Dark Energy halo surrounding the structure to keep it secure. It also has an Event horizon halo around the neutral center 0 point. We know that the black hole draws matter to itself and that when matter comes into contact with the center it is broken down. On route to the Event horizon the matter is gradually being stripped and decayed in a way that is similar to the way Neutrino heads decay.

From this information it is possible to theorise that in the core of Neutrinos, the matter approaches the Event horizon, Z Boson in a spiral and decays to the point where it crosses or quantum leaps into another state from (+) through (0) to (-)During this leap some energy is released because the impact should create a micro quantum boom where Neutrinos are emited into the cosmos. These Neutrinos go on to form new Quarks, that form atom which make up matter. In the case of the black hole, the matter had sped up on approach to the centre until it almost reached light speed. At point 0 there is nothing physical to slow the now subatomic matter down because it is traveling through a vacuum. It now outruns light speed.

Eienstien's theory of relativity supports this idea by saying that light is slowed whilst travelling through space because it is moving through particles. However if light were able to move through a vacuum as in a black hole it would travel faster breaking the light speed limit and thus some energy would be released causing a cosmic boom in the process. The cosmic boom releases the Gamma ray energy in which the Neutrinos are spewed out in a jet.

It is possible to theorise that, upon moving through the 0 point the rest of the matter finds itself in an environment of opposite charged particles identical to the ones that had once been physical matter. These particles had now decayed into Anti-matter particles. Science says that energy cannot cease to exist it can only change state. So when a dying galaxy, world, plant, animal or person clinicly dies their life force or mind energy must according to this law, have to go somewhere. Their physical atoms seem to eventually decay to this plane further adding to the (-) plane's mass.

The center of the Black hole has Dark Matter and Dark Energy passing through it. All possibly traveling at or faster than light speed. If Dark energy occupied the center in a steady stream whist passing through the vacuum, the stream of Dark Matter may have the ability to effect the gravity of the positive Matter.

If we consider the rules of entropy and super-symmetry we can theorise that the black hole structure is modeled from the blueprint of the Majorana particle. We would find that the Neutrino' gluon flux tube/W-& W+ Boson and micro black hole Z Boson body might work in a similar way.
We know Dark energy can effect the gravity of galaxies. Therefore it is possible to conclude that the very origins and blue print of a black hole came from the memory/programming passed on from the androgenous Primordial Father /Primordial Mother Majorana particle. It "evolved" through the generations of particles which turned into atoms such as Hydrogen, Deuterium and Helium until stars were formed in Nuclear fusion due to the mass number of these particles. Stars carried on the memory/programming /blueprint though all generations of all natural atomic structures. These atomic structures carried on the Majorana memory/ programming/ blueprint until all known elements were formed.

A limit may have been reached because all natural combinations had been developed. Eventually stars burn out leaving black holes. Some black holes are particularly noticeable because of their mass. These black holes give birth to new Neutrinos which may begin as baby/ micro Majorana particles that grow into Neutrinos almost instantaneously. The Neutrinos which go on to form Quarks that form the nuclei of the atoms, will go on to form the next stars. All based on the blueprint/ programming/memory of the evolving Majorana particle. This process can be thought of as an "ouroborus" cycle.

Another Dimension

It is possible that the mid point Z Boson/ 0 vaccum zone of the Majorana Neutrino serves another purpose apart from decaying and filtering matter through its micro black holes. It is possible that they all have a gravitational effect on each other inside the Nucleon. The mid points line up and are within reach of each other. They should create an energy force within the cube structure that forms a divide between our physical world and the dark matter world of "decay". Separated by the Z Boson Micro Black Hole system.

This field would behave like a veil of Dark Energy which light does not enter. Light only bends around it.
The upper corners of the structure seem to possess a (+) property while the lower corners seem to possess a (-) property.

The zones which are (+) would have a left-handed or clockwise net spin property of 1/2. The energy is existing at a lower state than its (-) counterpart. The energy in this space should be capable of forming physical matter in the universe.
Amino acids have left handed spin properties. If they had right handed spin properties they would not form and life as we know it would not exist.
The corners of the (-) zones form clear areas that are separated by the micro black holes/ Z Bosons which exist in the transition points between the Neutrino head and Anti-Neutrino tails. Each zone is made up in opposite quarters of the structure. The zones exist in a checkerboard pattern at a 90° angle from each other.

The zones which are (-) would have a Right handed or Anti-clockwise net spin property of 1/2. The energy is existing at a higher state than its (+) counterpart. The energy in this space should be capable of forming Anti-matter in the universe.

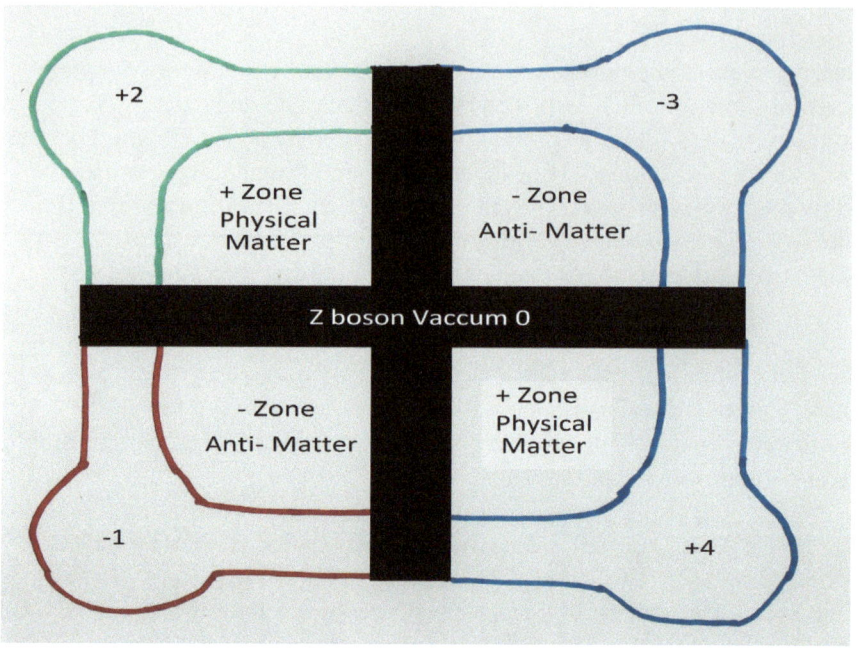

Cross section showing dimension divides.

(+) & (-) zones in the Deuterium atom.

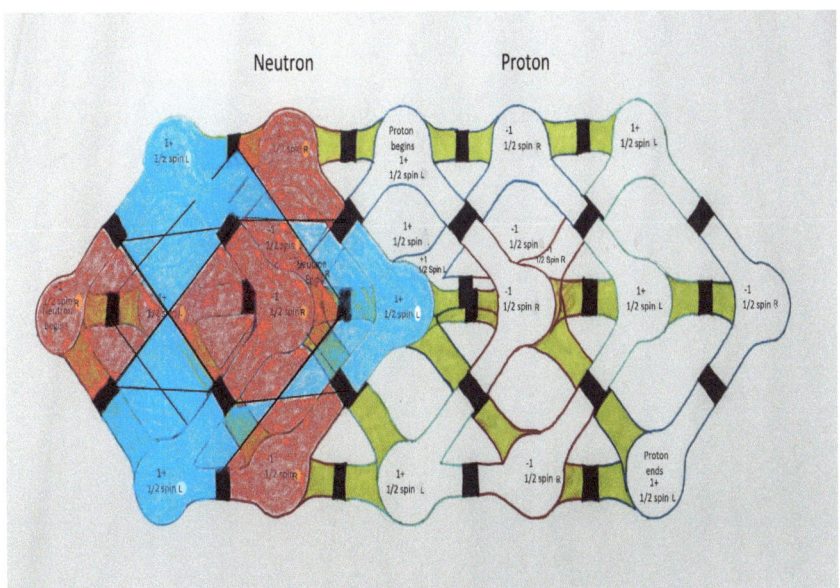

This mechanism should exist inside each atom of our bodies. This means that for every matter atom and cell there is an Anti-matter counterpart that exists inter-connected and entangled with our matter particles. If we think about this on a larger scale then we know that most of the space inside our atoms are made of seemingly empty space, in that we have failed to detect it with our instruments. This empty space is actualy the Anti-matter or dark-matter half of our existence. It would be an exact dark-matter copy of our matter self. It resides within our bodies cells.

Our universe can only be seen as a whole by taking into account the two half spins of both dimensions or planes of existence. The (+) plane consists of all of the energy and matter that we can perceive and detect with our technology.

This ranges from the planets to gasses and sound waves to human beings. The two dimensions are so closely inter-woven and entangled together at such a tiny sub atomic level that we can detect it only because we notice that something other than the (+) physical energy makes up the rest of the mass in our universe.

The (-) plane is undetectable to us because its properties lay beyond what our current technology is able to detect. This plane would consist of the Dark Matter and Dark Energy equivelent to Every Matter particle in current physical existence. However the size or mass of the (-) zone/ plane far out weighs the (+) physical planes mass. This is because the other (-) plane also contains Every particle that has Ever decayed from the physical plane over billions of years since the universe began.

So there is more Dark Energy & Dark Matter in existence than there is normal Matter & Energy at any one time.

Measurements of the X dimension:

It would be difficult to measure the dimension of decay because it is hard to detect. However we could try by taking into account the measurements that we already have for the physical plane. We could use the Hydrogen atom to do this.

The radius of Hyrdogen is 0.053nanometres. Its Proton radius is 2.1424^{-21} ferometres. The Nucleon-Deuteron model shows that the Proton consists of two positive up Quarks and a negative down Quark. Each Quark consits of positive up and negative down Majorana Neutrinos. The positive aspects give the Proton its physical qualities while the negative aspects give it its Anti-matter or Dark Matter qualities. When these areas are looked at in the cublike formation that the nucleus seems to form, we can see clear areas of positive and negative zones. These zones within the nucli of all atoms might be responsible for our entangled physical and Anti Matter dimensions to exist along side each other. They were created simulatiously along side each other and are forEver entangled. This is why we can detect that Anti Matter or Dark Matter is there but not interact with it as such. We are divded by these clear barriers facillitated by Z Bosons. The radius of the Proton is 2.1424×10^{-21}, if we divide this by two for the anti-particles we can see that each zone with positive matter potential and negative Anti-matter potential is half the size of the Proton.

$$\frac{2.1424 \times 10^{-21} \text{ ferometres}}{2}$$
$$= 1.10712 \times 10^{-2} \text{ ferometres}$$

so in the Proton of a Hydrogen atom the size of each dimension both positive Matter and negative Anti /Dark Matter should measure 1.10712×10^{-2} ferometres.
Both dimensions together = Space-Time.

For Deuterium which has both Proton and a Neutron:
2.1424×10^{-21}fm x2 (for both neurton and proton) = 4.2848×10^{-21}fm.
So Deutrerium length radius might be 4.2848×10^{-21}fm.
Proton size:
2.1424×10^{-21}fm
side:00000000000000000004241.2
vol:76289761678.528
surface area: 107926664.64
face diagonal: 5997.9626
space diagonal: 7345.9739

Deuteron:
4.2848x10^-21fm
side 000000000000000008482.4
vol: 610318093428.2239
surface area: 431706658.56
face diagonal: 11995.9251
space diagonal: 14691.9478

Matter dimension and Anti/Dark-matter dimension.
1.0712x10^-21
side 000000000000000002170.1
vol:10219725735.101
surface area: 28256004.06
face diagonal: 3068.9849
space diagonal: 3758.7235

Therefore it is also possible to conclude that this extra Dark Energy of decaying particles not only from black hole decay but the decay of all matter in general is the cause of the Parity symmetry violation (why Dark matter overthrew the production of matter in the early universe) and why the universe is expanding. Matter is being decayed faster than new matter is being produced.

This Ever-growing source of decaing Matter into Dark matter and Dark energy causes the universe to accelerate.

The production of the first Majorana & its Majorana Neutrino pair state to the production of the first Neutron seems to have been a state of equilibrium between the amounts of (+) & (-) energy & matter. It is possible to conclude that the decay of the first Primordial Father / Primordial Mother Neutrino /Anti-Neutrino pair caused an Ever-growing imbalance.

This first decay could also mark the begining of time passage as we know it because time is mearly the effect of change though decay. During this time the (-) zone is growing more massive as a result of these decayed particles enering it. These particles would now be at a state of equilibruim and no longer decaying. The two halfs unite to become whole again as (-) energy or matter.

A Spiritual Link?

The universe is estimated to consist of 68% Dark Energy, 27% Dark Matter and 5% Normal Matter. Psychics have described the world of "spirit" as seeming to be off to the right side of them. Some say that it seems to be slightly higher up than our own as though it were on a diagnal axis. This description seems fitting for the plane found in the Nucleon-Deuteron model.

For each (+) matter plane there is a (-) plane that exists to its right, meaning that it has a right handed Chairality. The state or frequency of the (-) plane is higher than anything that exists within our physical (+) plane and so it might be felt by a psychic to be higher and of a higher frequency than our own.

We know that the energy of particle decay has to go somewhere. We have worked out how this energy should decay to the point of entropy.
The next big question is to find out if there is any scientific evidence that intelligent consciousness can exist outside of the physical (+) body in its new (-) environment?
To do this we must open our minds and embark on a journey to the "dark side" of our consciousness, into the plane of possible disbelief. Where people have experiences that seem to defy science. This plane never-the less is very real and has been overlooked for far too long. Lets explore the (-) plane.
What would it be like to exist there? what are its properties?
Can intelligent life flourish there?

Science and Consciousness

If we apply what we have learnt from the Nucleon-Deuteron model to the human mind we would see that:
In the human brain, information enters the thalamus which is two egg shaped structures located at the back of the head. The thalamus sorts out information and sends it to the cortex. If the thalamus becomes damaged then consciousness is lost. So the thalamus might be the seat of the conscious mind.

As energy and matter in our physical plane began to develop into (+) consciousness, its counterpart particles would have also developed alongside it. Since the (-) seems to carry the information it might be this energy that held some of our innate programming just as the Majorana particle passed on innate information to its descendent particles.
The conscious mind might have developed in order to control bodily functions to ensure survival of any particular life form. The conscious mind is only concerned with the present moment and the task at hand. It receives and makes sense of current sensory perceptions that we usualy gather from the (+) physical plane. These perceptions and experiences each lasting only briefly are then processed and sent to the sub-conscious part of the mind for storage. Each experience lasts only briefly second by second. Each instant quickly decays to become the past, just a memory in the sub-conscious (-) zone of the mind. The sub-conscious mind contains all past "decayed" thoughts and experiences. When we need to access information from the sub-conscious storage system, our conscious mind contacts the Sub-Conscious mind that is responsible for storing the memories, which are thought to be stored in the brains microtubules. Microtubules are found in neurons and are known to fall apart with age casing alzheimers.
The brain is the physical while the mind is energy which flows through the brain.

Neurons fire in our brains, the neurons fire electrical energy which connect with each other via a network of connections. The information is retrieved from the sub-conscious mind and sent back into the conscious mind for examination. So when you remember what a loved one looks like it is not your conscious mind that knows this, it has to ask the subconscious mind for an image or memory of them. The (-) sub-conscious mind contains the information of experiences that make you who you are.

Our subatomic particles, atoms, molecules and bodies have a (-) Dark Matter counter part as does Every particle of matter in the universe. Our (+) conscious minds seem to have dark energy (-) sub-conscious counterparts.

This (-) sub-conscious mind communicates back and forth via Electro-magnetic signals along the neuron system. When the (+) conscious mind decays it would reach a state of entropy as do other particles and energy. Once the functioning of the physical conscious mind is no longer detectable by us because the person has achieved a state of physical & clinical death, we assume that the mind no longer functions. The brain needs the mind to make it function like a computer operator. If the operator or mind vacates the computer then the computer will cease to function. However the operator does not need the computer to exist. Just as the mind might be able to exist with out a brain. This is because the awareness of the first most primitive organisms may have evolved a brain to process experiences and to control increasingly more complex bodies. This would mean that consciousness created the brain. Once the brain dies the consciousness ceases to function on the vibrational level it was once accustomed to inside the physical brain. However we know that energy is never lost. All energy must go somewhere.

Upon the bodies death, the energy of our consciouses mind which was needed to function the body has now decayed into the (-) zone to become a whole with its (-) sub-conscious counter part. Now there is no separation of the two halves of the mind and so the mind as a whole can recall all past experiences instantaneously at will. The mind and thought patterns which hold the personality would be incapable of dying in the (-) plane, there is no need for decay because each particle is now one with itself. For this reason there would also be no measure of time as we know it. There just exists the constant "now".

The Energy and Dark energy inside the Nucleon of our atoms, which make up our DNA hold this energy or life force. The life force energy flows through the human body and brain.
Some cultures call this energy Chi or Prana. The life force energy of the human body is thought by Eastern cultures to run along the bodies electrical meridian system.

The model shows that within Every Quark there are Neutrino heads (+1) with the potential to evolve into physical matter of say a human body, and Anti-Neutrino tails (-1) that have the potential to evolve into Anti- matter or the seemingly empty space that makes up our bodies. This space is by no means empty. When atoms combine the mass of the Anti-matter increases along with the matter relative to the number of Neutrino heads inside the Quarks.

The brain may be physical matter made up using the same principle as the (+) Neutrino which lies dormant until it interacts with (-) or in the brains case brain waves / consciousness.

If the information carrying (-) Dark Energy & Dark Matter particles were taken out of the body, the body would be dormant like the Neutrino head. The body would cease to function = Death.

We know that thought is made up of Electro-Magnetic energy that runs along neurons in the brain.
The Nucleon-Deuteron model suggests that upon the death of the physical brain the energy does not cease to exist but transforms. On a subatomic level It travels from the dying brain at close to light speed through the gluon flux tube/ W-& W+ Boson and Z Boson/ black hole structure during its process of decay from the physical dimension. It arrives in the (-) dimension of decay where it becomes energy there.
Therefore any person who's consciousness becomes completely separated from the body upon death should find their consciousness existing in a new dimension. Their consciousness would now be existing at a state that is faster than light speed and existing in a "timeless" reality that has different quantum rules to our own.

If we look at Dark Energy and Dark Matter's role in creation, we can see that when the (-) which carries information & mass merge with (+) matter they create life and functioning of the body and brain, rather than annihilation. This is why our bodies are made up of seemingly empty space. This space is our "Dark side".

We know that upon death the memory or consciousness seems to withdraw from body.
People who have had near death experiences have reported seeing their whole life flash before them to be viewed in real time. This is possible if upon the death of the brain the (+) conscious mind is merged with the sub-conscious mind. All past memories and experiences would at this point become apparent to the person because the entirety of both areas of the mind are linked and in communication with each other. They become one as they exit the brain. The dark tunnel with the light at the end sounds as though the persons consciousness were traveling toward the Event horizon of a black hole. However this same vision would be experienced if the consciousness could observe travelling along a gluon flux tube/ W-& W+ Boson structure. Although some people have big personalities the consciousness itself might be quite small. It is energy which travels upon tiny neurons in the brain, it must be quite capable of travelling along the gluon flux tube/ W-& W+ Boson structure of the sub-atomic environment.

Within an instant possibly at near light speed, people have reported that their consciousness as hovering outside of their physical body. The subjects later describe the Events that took place whilst they were being revived at a point when they had been declared clinically dead. Some scientists suggest that these experiences happen inside of the dying brain and that they are tricks of the mind. This could be a reasonable explanation if it wasn't for the fact that these subjects report floating outside of the body looking down from above. If this were all just an illusion of a dying brain how did the subjects eyes get outside of the body for an accurate viewing? The only explanation for this phenomena is that the subjects must be viewing the scene from an alternate angle and retain the ability to think and make conscious observations and decisions whilst outside of their body, during the bodies state of clinical death. Their physical body has ceased to function and is laying dormant. The empty body cannot process thoughts because the brain is unresponsive and no neurons are firing. The heart has stopped and so no oxygen is able to reach the organs to allow any manner of functioning. The eyes are generally closed and tend to still be inside the skull, unable to observe any angle what so ever.

One instance records a lady who had been born blind. She could not describe how anything looked and could not describe colour. During her near death experience she recalls hovering outside of her body looking down at herself, as she was being revived. She saw people (Doctors) for the first time and saw colour. She saw what the objects that she had only felt her whole life looked like. When she was revived she described these experiences but remained blind for the duration of her existence inside her physical body. This experience could only have been possible if her consciousness had genuinely been outside of her body.

Scientific experiments were carried out on genuine mediums during trance and psychic activity. The experiments mapped the brainwaves of the genuine mediums. Results found that while genuine mediums were obtaining information from the minds of the so called "deceased" person. Their brainwave function showed Delta - Theta frequencies.
This is when they are obtaining information from the other dimension or spirit relm.

Mediums have reported that the other dimension of spirit is timeless. They do not experience time as we do. We know from entropy that we measure time through change and decay. The law of relativity would agree that if the parallel plane exists past the (0) transition point on the negative side, it should exist in a state that was so fast that it would "outrun" light and time would cease to exist. There would be no decay because particles have already decayed to their fullest and are now existing in a state of equilibrium. Beings would therefore be unable to experience ageing, illness or death. They would remain in a state of a constant "now".

Time is irrelevant because they have no decay to measure it by. The only way they might be able to measure time is by observing the newly decayed matter which is causing both the Dark Matter plane and our own matter plane to expand and accelerate.

From this information it is possible to conclude that there is only one other parallel unseen existence, and that the universe seems more of an evolved particle than hologram, computer game, or multiverse. There exists only one other half of unity. Together these two dimensions add up to the unity of the spin properties of the subatomic particles, and also complies with the law of supersymmetry.
They comply with the theory of relativity and explain the missing mass together with the mystery of quantum entanglement of particles. It also offers a solution to the split photon mystery.
While there seems to be only one other 'dimension'. That dimention is most likely divided into frequency 'zones'.

What Properties Might The Parallel Dimension Have?

Once particles including the mind energy have decayed from our physical dimension into the parallel dimension of "decay". They might find that they are a perfect replica of their previous particle state but opposite in nature. Therefore the parallel dimension could be constructed from Anti-Neutrinos with right handed charility that make up the down portions of the Quarks and previously decayed matter & energy. These combined make up the "missing" space that seems to be inside our atoms and in the universe in general. These Anti-particles have more mass than normal particles.

When energy passes into the parallel dimension it might behave in a similar way to the way it did in the physical dimension except its rules of physics might be opposite to our own, to a certain degree. In other ways they might be able to do things that quantum laws on our side will not permit because we travel slower than light speed and are operating on a frequency which is far too slow. There the different laws of physics might enable a being to defy gravity enough to fly.

Particles might be free to move around or fuse with other particles. If their building blocks of matter work in a similar way and our Anti- particles make up a copy in the X or Spirit dimension, then a (+) physical door should have its (-) Anti-particle equivalent. This equivalent might be closely inter-twined with the physical door. Therefore the parallel dimension would also have a copy of the door existing in it. The parallel dimension should contain not only already decayed particles in a stable state but also the Anti-particle half of Every particle in existence. This would mean that the human body has a physical (-) Anti-particle body made of physical matter that we can measure. We know that we are also made up of seemingly, mostly "nothing", "empty space". This space is not empty but is made of the Anti matter/ Dark matter particles that make up the body. In the parallel dimension they should make up a dark matter copy almost like a shadow of the body. This copy is the physical bodies (-) counterpart. It is the body which the consciousness would be focused in while it is existing in the (-) plane.

This Anti-particle body should stay entangled, with the physical body until the bodies physical death at which point it should travel over to exist in the (-) plane. This could be why Psychic mediums report that when they see spirit beings in their minds eye or in apparition they look the same as they did in "life", they are said to retain all memories, love, mannerisms, personalities and consciousness as they did when they were incarnated in the physical dimension. Particles and beings in the the parallel dimension are thought to have the ability to pass through your physical matter. This might be because our sub-atomic structure is opposite in nature.

Take A Walk On The Not So Dark Side

Dark matter is called "Dark" because in scientific terms we cannot percieve it with the naked eye or with instruments. It was not thought to interact with matter. Light seems to "bend" around it, it also effects gravity. It is thought of as a mysterious substance that prevades all matter in the universe. However Psychic mediums and people who have had NDE's have reported that the spirit dimension appears to them as full of light. They describe an all prevading light of the highest frequencies that glows and radiates warmth it makes them feel good. This sounds simaler to the way we feel when we sunbathe. They describe seeing colours and hughs that they have never seen before and cannot describe. This indicates that they exist in frequencies that might exceed our known spectrum. As we know their dimension contains Everything that as decayed from the tiniest particle to all decayed stars. The amount of decayed stars that have carried their energy into this dimension since the beginning of the univesre must be immense. This could be the source of their all pervading light. It should be far from Dark there.

The light of this parallel dimension is known to the psychic community as source. According to psychic mediums and those who have claimed to have witnessed it during NDE's where clinical death was recorded, The source is beleved to be far from a simple energy. It is not dormant but contains and pervades all that is in the parralel (-) dimension and the dimension of matter. It contains all informaion and all knowlege. It seems to have controlled the evolution of the universe and caused consciousness to evolve with it.

It is the culmination of everything that was (now decayed matter), everything that is, in both dimensions and it will facilitate everything that ever will be in both dimensions.
Our thought waves become part of this matrix. This is what Psychic mediums talk about when they refer to the Akashic records. The Akashic records are thought to be a certain vibrational frequency within the parralel dimention that "spirit" beings and psychic mediums can "tap into" or tune into, in order to find out information. It seems like it is a sort of natural internet system or cosmic mind.

Consciousness Evolution:
Where We Came From & Where Were At

Consciousness seems to have developed from the self preservation needs of subatomic particles. They developed into atoms which found themselves in different environments in which they met with new needs. They developed more self preservation skills and passed them on to future generations though programming or "innate memory". Atoms turned into matter which turned into planets. Our planet was created and the elements developed, from this amino acids and cells developed. The cells turned into simple life forms, these simple life forms reproduced themselves by splitting to produce a compleate copy of itself. The first life forms were androgenous. They developed yet more needs for self preservation in ever changing harsh environments of the early earth.

It would have been energy consuming for a lifeform to reproduce and find nurishment especially if it had to move to get this. Life-forms developed a method to conserve energy. The androgynous life forms developed a way to split into two forms each with specific functions. Because they both served different functions, they developed different body types that were appropriate for their function. These two different forms became the two sexes. Females would have used their energy to reproduce and developed a womb for this job. The male lifeform would find nurishment and provide protection for them both and their offspring. This is why a lot of male species are built physically stronger or larger than the females. The males also developed the ability to produce sperm.

Most modern day humans no longer have to hunt wild animals for tea or fend off wild animals. So the males are free to do other jobs. In our society we have childcare facilities so parents both male and female can be free to work and live differently to how we might have in the stone age. Consciousness has evolved lifeforms through the ages thanks to DNA which also seems to carry information programmed into it from the Majorana first particle.

Consioussness has evolved in humans to become self aware enough for us to realise that we can be free to live the way we want to. Most of us no longer have to be slaves to our environments quite so much. This freedom of self aweamess allows us to express ourselves in what Ever manner we wish.

What Can We Understand From The Nucleon-Deuteron Model?

What does the Nucleon-Deuteron Model reveal about the Photon?

The Nucleon-Deuteron model explaines that the Gluon flux tube / W-& W+ Bosons / Z Boson/ black hole system is not only responsible for providing mass and strong force inside the Nucleon which hold the Quarks together. They are also the medium through which information is passed or transmitted between Quarks and the Neutrinos. The essence or "innate memory" of these interactions are still found in the Photon.

The Photon travels at light speed and its Anti-Photon should travel faster than the Photon breaking light speed because it exists past the (0) point. The Anti-Photon exists in the (-) state or dimension in a faster state. This frequency might range from Theta and beyond. Because of the nature of the Anti-Photon and the Photon and their entanglement via their own Gluon flux tube / W-& W+ Boson and Z Boson black hole connection they should experience instant rapport betwween them.

The Gluon Flux tube/ W-& W+ Boson/ and Micro black hole system does not diminish with distance, it doesn't matter how far the particle seems to be from its Anti-particle they remain entangled. The Gluon fluxtube/ W-& W+ Boson is capabel of stretching to facilitate changes of distance, however the diameter stays the same. Therefore the same amount of information can be passed through it between Neutrino, Electron and Photon head & tail ends.

What Does This Mean For Our Future?

These answers might help us better understand the inner workings of the atomic nucleus which could lead to a better understanding of the universe as a whole. A better understanding of Dark Energy & Dark matter could lead to more advanced technology that could revolutionize the way we detect and fight diseases & Illness.

It is possible that we might also be able to gain a better understanding of the ageing process. Until now science has understood the ageing process as being due to a body clock which shuts down cells after they perform a certain number of divisions. The Nucleon-Deuteron model suggests that ageing is caused by the energy flow along the W Bosons being drawn toward the Z Boson. During this time a cell which is on a much larger scale will manage to achieve a certain number of divisions until it's Majorana Neutrinos finally reach their Z Boson event horizons. This should ultimately lead to decay. Along its journey the cell loses energy and is less able to reproduce itself as effectively causing mistakes or cell mutations. This might again be due to the Neutrino energy being striped on route to its Z Boson event horizon. What happens on the smallest atomic scale effects the cells on a larger scale. This might help to explain why cells divide with mistakes or mutations that can cause tumours. The damage or decay seems to be happening to the Neutrinos as they lose energy from (+) to (-) the energy loss might also cause loss of information of the Neutrino, which effects the atoms that effect the amino acids, which effect DNA and replication of cells. Thus causing the cell to lose it's ability to create a perfect copy of itself. Some cells could reach a point where they do not have the correct information to do the job perfectly thus imperfect cells are created.

If it were possible to pull the decaying energy back up though the gluon W Bosons away from the (-) zone and the event horizon, we might be able to keep ageing at bay and ensure that Neutrinos remain in an optimal "state" so that they can perfectly replicate. This might result in perfect cell divisions and longer life spans. This might only be possible if we could somehow use energy to draw the (+) energy back up the W Bosons. This might not be as strange as it sounds. Reiki and other spiritual healing practices seem to work on similar principles. The practitioner uses their own energy (Chi / Prana) to manipulate the patient's energy back from disorder to order. Or in the terms of spiritual healing from dis-ease to ease.

A better understanding of Dark Energy and Dark Matter could also lead to communication between the two parallel worlds or dimensions that make up the entirety of the universe. At the moment the only technology that exists for this type of communication is EVP recorders and infra red detectors.

It might be possible to develop a quantum binary transmitter receiver device that uses frequencies of Delta or Theta. A quantum radio transmitter or phone device might be able to send and receive messages from our dimension directly to this astral/ parallel plane. It would provide an easier more accessible communication available to none psychic people.

Maybe one day science will be able to re-contact those great scientific minds of the "past" It is possible that the thoughts of a disembodied conscious minds who now exists in the dimension of decay, may be able to send their thoughts as messages into the minds of an incarnated physical mind or purpose built machine. Psychic phenomena is extremely difficult to prove because these processes occur inside the mind of a medium. However ground breaking experiments have produced much data that provides evidence that something strange is going on inside the brains of the mediums during this claimed spirit contact. Up until recently science has not recognized this phenomena because it had no means of measuring it. However just because something cannot easily be measured does not necessarily mean it does not exist. This has been proved many times over with the prediction of subatomic particles that were indeed found many years later. Also with the existence of Dark Matter & Dark Energy. It is therefore worth examining the scientific evidence that has been collected so far of such phenomena.

Psychic Mediums Claim To Communicate With The "Deceased" But How Are They Doing This?

Some Psychic mediums communicate using Delta & Theta frequencies. The brain uses Delta frequencies when it is in a deep unconscious state or deep sleep. Most mediums meditate regulaly to keep their minds frequency high enough for communication to occur. Their brainwave frequencies have been scientificly measured at delta- theta.

For the thoughts of a disembodied mind to reach an incarnet mind the waves might be being transmitted back up through the gluon flux tube/ W-& W+ Boson system to meet the receivers in the recipient mediums mind on the (-)Event horizion. In theory if thought frequencies were passed back up through the structure they would travel near light speed. The thoughts would be registered in the mediumms mind as fleeting thoughts at the lowest frquencies of dalta, or theta. The waves once recived would decay almost instantly back through the gluon flux tube W-& W+ Boson structure to the dimension of decay. The medium is left with the memory of fleeting images or words that are remembered and passed on to the sitter or person for whom the message was intended. Most people are busy and do not tend to meditate, they exist by mainly using Alpha brain waves. This is one reason why most people are not sensitive enough to receive this communication which requires the use of these frequencies. Most people think that they are not "psychic" but in fact they could also tune into these frequencies if they trained their minds to do so. Because of this many believe that mediums cannot communicate with spirit beings, because they themselves cannot.

Spooky Healing At A Distance

These same delta and theta frequencies have been observed in the brains of psychic mediums & energy healers while they are sending spirirual healing. Mediums envolved in scientific experiments by neurologists had their brain waves traced. It was found that the neurons at the back of their brains were firing more than when they were not in a healing state. They fired more than an average persons would. Different parts of their brains fired until the energy trail of neurons firing entered the frontal cortex of the brain which is responsible for compassion. The energy appeared to shoot out of the brain. The subject who was being distantly healed was also monitored. It was found that at the time the medium had sent healing the recipient has almost identical nuron firing inside her brain. The recipient had a dramatic reduction of Alpha waves which occur in the normal wakeful conscious state.

People who recieve this type of healing including Reki seem to experience great relief from their ailments after the healing.
We know that we are made of energy. Spiritual healers believe that we would not be alive if it were not for the electric current/ Chi /Prana we have running through our bodies. If this current is not circulating properly around the body the energy is disrupted and not at ease. This equils "dis-ease" of the energy, which leads to illness. Spiritual healers beleve that a person can effect and create balance in anothers electical meriden circuit by using their own Elecro Magnetic energy to redirect the patients energy flow. This leads to the ease of the energy and wellbeing. Many people have claimed to have been healed at a distance. Is this "spooky healing at a distance" and could be facilitated at an atomic level by the Gluon Flux tube/W-& W+ Boson, structure that could also be responsible for the ageing process.

If this is so, it should mean that our minds are entangled with space-time. Space-time contains Neutrinos of both Matter and Dark matter, so do our brains and the neurons that fire including the thought energy itself.

Some mediums do not necessarily need to go into trance, they can communicate by just shifting their attention slightly.
More research is needed into brain wave frquencies of practicing mediums.

Accessing The Parallel Dimension

The Nucleon-Deuteron model also revealed that there are at least two ways that it would be hypothetically possible to access the parralel dimension. It would mean opening a worm hole which according to the model might be possible by making use of the Gluon flux tube/ W-& W+ Boson Structure, and by making use of the spectrum in particular ways. This would be highly dangerous as we run the risk of opening a hole that we cannot close. In a hypothetical worm hole, only information might be able to pass through. For our physical bodies it would unfortunately mean Death.

Reversing The Ageing Process

From the Nucleon-Deuteron model we can gain a greater understanding of how the universe was created and the mysteries of life, decay and death. With this information, science might one day be able to develop a tecnique to slow the ageing process. We might also gain more knowledge concerning Dark Matter & Dark Energy for use in early detection of disease & illness.

The model offers science a new direction and opens up new possibilities for exploration.

Final Thoughts

We are one, United we stand, Divided we fall - lets stand tall!

We're all created from the same energy. We all "decay" to the same parallel dimension regardless of our beliefs in the "afterlife" or religious beliefs and preferences.
The universe has fought since the begining of time for its own self preservation from the very first Majorana particle that produced a protective membrane around itself, changed form and replicated to carry on its energy.

This protective nature through consciousness has developed and evolved into compassion and love for one another in both dimensions. The parallel dimension has no decay, illness or death. There is no need for our survival techniques which lead to greed and a number of "deadly" sins.

People in our dimension seem to have forgotten that we all come from the same source, we are all neighbours on our tiny, little planet. We all descend from the same common ancestor and are therefore related. Some people have developed light skin over time as some of humanity moved away from hotter climates, where dark skin was necessary for sun protection. The melatonin in the skin block harmful uv rays from the sun. This is okay if you live in a hot climate because the intence sun also delivers much needed vitamin D to the body that is absorbed through the skin. Once a person moves into a colder climate where the suns energy cannot be felt as much, there is less need for UV protection. However there is a large need for vitamin D which cannot be absorbed as much in cooler climates. To combat this problem, people who settled in cooler climates began to loose their melatonin in the skin over time through evolutionary processes, thier skin became lighter. This was purley because lighter skin absorbs essentual vitamin D more effectively because it does not block UV rays as much as dark skin does. So the difference in our skin colour, does not mean that we are different it just means that some people live in different places and have developed different ways of self preservation because of dietry needs. After knowing this information there is no place in the world for racism or persecution of any kind based upon physical differences.

Our world is the only dimension where we seem to enjoy focusing on our insignificant minor differences. We use them to segregate each other and to discriminate against one another. In ancient times religion was understood in a simaler way all accross the known world. This is why most religions have a very similar creation stories, morals, angels, etc. Ancient mystery schools would teach kings and the greatest minds from all over the known world. They would all learn the ancient arts of masonary, science, geometry, astrology and religion. They shared their knowledge & ideas and took them to their home countries to be utilised and shared. This is why simalar ancient monuments and ancient religous ideas can be found all over the world.

However over time, because of territorial wars we have become more segregated and become more out of touch with each other. This led the human race to become isolated from each other and religious beliefs became more developed in different communities. False ideas of separateness began to dominate our human race until now we find that we can no longer talk about religion or the nature of a Primordial Father or God for fear of offending someone. Each religion believe that they are the chosen favourates of their God. The belief that a God segregates and discriminates against all others who do not believe in what they do, has led people into numerous wars through-out history.

To avoid repeating past mistakes we must learn from our past as a united world. We must realise that most of our religions say very, very similar things to each other because they were all derived from the collective knowledge and understandings of the ancients. Their ideas of a creator were simaler accross the Ancient known world. Differences in religion these days arise from teachings which have been known to branch off into entirely different religions because of the smallest of facts. The truth of the matter is they both believe in the same divine creator.

In the parralel dimension we are all the same. There is no segregation, or descrimination because this fundamental truth is understood there. We are all equal to each other regardless of our religious beliefs. To kill and descriminate in the name of God violates the protective, moral, intelligent nature of the universe and the divine collective creator force. Because such behaviour is a purely earthly, primitive human thing to do. It shows that the individual who displays such behaviour has not fully grasped the concept of the true nature of the divine.

As we know the parallel dimension is far more superior and intelligent than our own with its vast knowledge and understanding of the true nature of our being. Prejudice, and discrimination are man made ideas that are considered wrong and have no place in the parralel dimension and are not encouraged nor understood there. These behaviours stem from the thoughts of man. The all protective, all accepting creation force, the energy of all that is, does not order such atrosities! So racial and religious descrimination has no place in this life or the afterlife!

Our world could progress if we stop trying to kill each other, making powerful bombs that might one day destroy our planet. True progression would be if we could utilise our knowledge to create cures for diseases and tecnology to better our world.

You do not have to beleave in a conscious Spirit world to believe in this Model, you just have to beleave in the laws of physics to understand that this is highly probable!

Peace be with you.